SUGAR

SUGAR

CHEMICAL, BIOLOGICAL AND NUTRITIONAL ASPECTS OF SUCROSE

Edited by

JOHN YUDKIN *Professor of Nutrition*
JACK EDELMAN *Professor of Botany*
LESLIE HOUGH *Professor of Chemistry*
Queen Elizabeth College

LONDON BUTTERWORTHS

THE BUTTERWORTH GROUP

ENGLAND

Butterworth & Co (Publishers) Ltd
London: 88 Kingsway WC2B 6AB

AUSTRALIA

Butterworth & Co (Australia) Ltd
Sydney: 20 Loftus Street
Melbourne: 343 Little Collins Street
Brisbane: 240 Queen Street

CANADA

Butterworth & Co (Canada) Ltd
Toronto: 14 Curity Avenue, 374

NEW ZEALAND

Butterworth & Co (New Zealand) Ltd
Wellington: 49/51 Ballance Street
Auckland: 35 High Street

SOUTH AFRICA

Butterworth & Co (South Africa) (Pty) Ltd
Durban: 33/35 Beach Grove

First published 1971

ISBN 0 408 70172 2

Filmset by Photoprint Plates, Rayleigh, Essex

Printed in England by Richard Clay, Bungay, Suffolk

CONTENTS

III SUCROSE IN THE PLANT

IV THE ABSORPTION AND METABOLISM OF SUCROSE IN ANIMALS

V SUCROSE AND HUMAN DISEASE

PREFACE

Sucrose is a unique substance of great versatility and commercial importance. It is consequently of interest to a wide range of specialists including the chemist, the biochemist, the animal and plant physiologist, the clinician, the historian, the sociologist, the nutritionist, the psychologist and the food manufacturer. Several of these interests are strongly represented in the research activities of the departments of Biology, Chemistry and Nutrition at Queen Elizabeth College. Because of our desire to strengthen inter-disciplinary studies at Queen Elizabeth College, we invited a group of scientists, whom we knew also to be interested in the various aspects of sucrose, to speak at a symposium at the College held in September 1969. These papers are now presented together here in the hope that they will be appreciated and enjoyed by many more people than the two hundred or so who were able to come to the symposium.

We were fortunate in having Professor Sir Edmund Hirst with us throughout the symposium and especially in having him give us, at the end of our discussions, a wide-ranging survey of the papers that had been presented. With his permission, we have printed his summary as an Introduction, so that readers may have an opportunity of seeing, in a few pages, the scope of what is covered in the more detailed papers.

Our objective was to cover the ground in such a way that each topic was reasonably comprehensible and comprehensive. In particular, we aimed to give information in such a way that it could be understood and used for reference by the specialist in fields outside his own particular speciality. We hope that we have succeeded.

JOHN YUDKIN
JACK EDELMAN
LESLIE HOUGH

LONDON

INTRODUCTION

SIR EDMUND HIRST

Emeritus Professor of Organic Chemistry, University of Edinburgh

The twenty papers presented at this conference and the detailed discussions which followed contain so much material and cover such a wide variety of important topics that anything in the way of a summary of the proceedings would be quite impracticable. All that can be attempted therefore is to record some comments and impressions gained during the meeting concerning the present state of the wide range of scientific investigations involving sucrose. The first comment must clearly take the form of congratulations to the organisers of this conference for their success in providing a programme which has emphasised so well the astonishing and unique role which is played by sucrose in so many and so widely varied aspects of chemistry, biology and medicine.

Fundamentally these activities are all related to the chemical properties of the sucrose molecule and it would seem appropriate to comment first on the group of papers which deal specifically with problems of sucrose chemistry and the utilisation of sucrose in chemical industry *(Chapters 5, 6, 7, 8 and 9)*. In this connection it may be said that sucrose investigations have always been difficult and the results often surprising, but frequently all the more important in consequence. The elucidation of the furanose structure of the fructose unit in the molecule presented major difficulties; moreover, the discovery that the five membered ring, contrary to all expectations, was not flat, had important repercussions on ideas of conformation which are so often used today in dealing with the reactivities of sugar molecules. The extreme ease of acid hydrolysis presents a major difficulty in the preparation of derivatives of sucrose and I think many of us were surprised to find evidence of alkaline hydrolysis also.

The presence of no less than eight hydroxyl groups in an awkwardly shaped molecule adds further complications to the problem of making derivatives of known molecular structure and emphasises the comment by Professor Hough *(Chapter 5)* that 'a fundamental knowledge of the chemistry of the sucrose molecule is widely accepted as one of the most difficult and challenging topics in the carbohydrate domain'. Nevertheless, as he goes on to say, such a knowledge is gradually emerging, and the examples from his own work and from those of other investigators given in his paper show how the difficulties are being overcome.

A particularly important aspect of this type of investigation is indicated in Professor Buchanan's paper *(Chapter 8)*. It is clear from other communications also that derivatives of sucrose, for example the phosphate esters dealt with by Professor Buchanan, are of fundamental importance in plant biochemistry and may have implications in medicine. Structural investigations and synthetic work on these substances are now pressing problems and Professor Buchanan shows us how modern methods making use of novel and ingenious techniques can be applied successfully in such syntheses. All this prompts the comment that there is now, and will be for a long time to come, an important role for synthetic organic chemistry working in conjunction with biochemistry and biology. The wide range of sucrose chemistry is further shown in *Chapter 9*, in which we are led into the extremely complicated reactions involved in the formation of coloured materials by the action of alkali on sucrose.

Finally, in this section, a comment may be appropriate on the use of sugar as a raw material for chemical industry. Some years ago in view of its availability and cheapness it was thought that extensive developments were likely, but, for some reason, little followed in the way of large-scale activity. More recently, however, interest has revived following the development of novel techniques and the availability of new solvents permitting reactions in non-aqueous media. Professor Hough gave some examples of this and Dr. Hickson *(Chapter 6)* has summarised recent progress in the utilisation of sucrose as a source of surfactants, polymers, plastics and nutritional supplements. At the moment, however, these are developments described by Dr. Hickson as likely areas for future activity rather than as established outlets for sucrose. One definite development was, however, described in detail in *Chapter 7* which deals with the successful use of sucrose as an ingredient in melamine resins. In this connection, the close interplay of all the varied aspects of sucrose science comes to the fore. If industrial uses of sucrose continue to expand and become really large-scale, decisions will clearly have to be made regarding the allocation of sucrose between its roles as a food and as an industrial raw material.

The next group of papers on which I would like to comment are found in *Chapters 10, 11, 12 and 13,* which are concerned with the role of sucrose in plant life. It is indeed an astonishing story. In very many ways sucrose here occupies a unique position, but it is evident from the present state of knowledge that there is no satisfactory explanation as to why this should be so. Both what has been established and the immense gaps in our knowledge are equally surprising. Sucrose is formed at a very early stage following the primary processes of the photosynthetic cycle and therefore has a predominant part to play in the life of the plant, both as temporary storage material, and as the medium by which the carbon and the energy necessary for growth and metabolic activities are translocated from one part of the plant to another. In these respects sucrose is pre-eminent, and even in the relatively few cases where other sugar residues are involved in translocation they are attached to a sucrose molecule. In many polymeric reserve materials stored in plants, such as inulin in dahlia tubers and levans in grasses, the form adopted in the polymer involves a terminal sucrose residue on which the rest of the molecule is built up. Yet it appears that we know far too little about the mechanism by which sucrose is synthesised in the plant and about the various changes it undergoes in the plant.

Chapters 10 to 13 by Edelman, Walker, Spanner and Fewkes give a fascinating account of some of the activities of sucrose in the plant and show how much there is still to be learned concerning the mechanism of primary formation of sucrose—probably via UDPG and fructose phosphate—and where it takes place in the plant. It has still to be established with certainty whether or not sucrose synthesis takes place inside the chloroplast system, and there are unsolved mysteries concerning the mode of utilisation of sucrose in the plant after the primary photosynthetic sucrose has been produced. It seems reasonably clear that sucrose during translocation moves from one site to another through the sieve tubes of the phloem and here we come up against the perennial problem of the botanists concerning the driving force which moves the sucrose solution. The movement appears to be physically and mathematically impossible—yet it takes place very readily! Dr. Spanner reviews the various suggested mechanisms and on balance inclines to the view that electro-osmosis is involved, depending upon the known high potassium content of the sieve tubes.

In all these investigations a major experimental difficulty is encountered in that complex enzyme systems are involved, and the observed state of affairs is the result of a delicate balance which is very readily altered and usually with disastrous results. Some indications of the complexity of these control systems is given in *Chapter 13* in which Dr. Fewkes discusses some aspects of the effects of plant growth hormones such as auxins and gibberellins in regulating the enzyme systems that promote growth and the accumulation of sucrose in the sugar cane.

Turning now to problems arising out of the use of sucrose as a foodstuff, *Chapters 1, 2, 3 and 4* deal with such matters as the relative proportions of sucrose and other carbohydrates in the human diet, the vast increase in sugar consumption per head in the western world, and the steps taken by the refiners to produce the numerous grades of sugar needed to meet the demand. Nearly all of us would agree, I feel sure, that the sweetness imparted by sugar to many articles of our diet is an attraction appreciated not only by human beings but by many animals also. It must nevertheless have come as a surprise to be told that in Britain the consumption of sucrose per head of the population has now risen to some 50 kg per annum and that almost one-fifth of the total calorie value of our food is provided by sucrose. It is of interest that in both the developed and underdeveloped countries the total carbohydrate in the diet is very similar in amount, but the richer nations use a greater proportion of sucrose. Whether this is good for the inhabitants of the well-to-do countries is one of the burning questions of the day and we have several chapters centred round these problems.

Some of these *(Chapters 14, 15, 16 and 17)* deal with biochemical problems involved in the metabolism of sucrose in animals. Dr. Bender and Mrs. Damji have provided a comprehensive review of our knowledge of the effects of sucrose, starch and glucose on various metabolic activities in man, rats and chicks. The results are extremely complicated and clearly depend on a large number of factors, but they do demonstrate that uncontrolled substitution of starch by sucrose can on occasions result in serious abnormalities. Dr. Widdas deals in detail with the mechanism of absorption of sucrose through the walls of the intestine. This takes place apparently after hydrolysis of the disaccharide and many of the problems appear to arise from the difference between the rapid active absorption of glucose and the slower passive absorption of fructose. Fructose absorption is relatively slow, but once absorbed it can readily enter metabolic channels in various ways, including isomerisation to glucose.

According to views expressed by Dr. Naismith, fructose as such is metabolised only to a limited extent by tissues other than the liver and it is suggested that the increase in fat concentration following this ingestion of sucrose-rich diets may be traced to increased synthesis of triglycerides in the liver followed by their transfer to adipose tissue.

The situation however is extremely complicated, which is not perhaps surprising in view of the very large number of enzyme systems already known to be involved in the metabolic processes concerned with the utilisation of sucrose. Additional evidence of this is revealed by Professor Macdonald. A further problem is that many of these physiological and biochemical investigations must necessarily make use of experiments with animals, and great caution is necessary in using the results as indications of similar activities in other animals including man. Indeed men, like other animals, differ very much from one individual to another, giving rise to further complications and rendering it very difficult to put forward definite generalisations.

All this leads to consideration of a group of papers which are of special medical interest. *Chapters 18, 19 and 20* are concerned with the role of sucrose in human disease. The first of the three by Dr. Valerie Burke deals with sugar intolerance, particularly in young children. The trouble arises by reason of enzyme deficiencies, for instance lack of sucrase, which interfere with the normal metabolic equilibria. Much is already known of the enzymic conditions which result in malabsorption of sucrose, glucose, galactose and fractose, and it is very satisfactory to learn that clinical procedures can be devised to cope with the trouble in many cases.

There is widespread interest in the prevention of dental caries, a disease which, as Professor Hartles records, seems to increase with improvements in living standards. Here is one instance in which sucrose appears to be definitely implicated, albeit in a somewhat roundabout way. In addition to a susceptible or vulnerable tooth surface, it is necessary for the development of caries that there should be present suitable micro-organisms and fermentable substrates. The course of events appears to involve the production of a polysaccharide from sucrose which adheres to the tooth surface and is then broken down by other micro-organisms to form acids leading to the destruction of the tooth. Sucrose is not unique as a sugar from which the polysaccharide plaque can be formed, but it is apparently particularly active, and in view of the large consumption of sugar, it may play a major part in the origin of much of the dental decay which is now unfortunately so common.

Finally we come to the important paper by Professor Yudkin in which he discusses critically the evidence that a high consumption of sucrose may be a contributory factor in the origin of many human diseases including dental caries, obesity, coronary thrombosis, diabetes and a number of diseases of the skin, digestive tract and joints. The indictment is a formidable one, but Professor Yudkin points out the difficulties there are in obtaining evidence for or against a correlation between diet and a specific disease, emphasising the caution necessary in the interpretation of statistical evidence and in the interpretation of experiments with animals. These difficulties are illustrated in a detailed account of studies on the relationship between carbohydrate-rich diets and a tendency towards obesity.

In the case of coronary heart disease, population studies and experimental results are providing evidence, mostly indirect, of a definite statistical connection between the incidence of coronary thrombosis and excessive use of sugar in the diet. Consumption of fats may also come into the picture. The mechanisms are still largely obscure and it seems possible that changes in insulin level and in the tendency of blood platelets to adhere to each other may be as significant as the widely canvassed rise in the cholesterol level. In respect of several other diseases, for example certain types of dermatitis, gastric and duodenal ulcers and gall-bladder trouble, the evidence against a high sucrose content in the diet is at present less convincing but cannot safely be ignored.

The problems in the biological field and in medicine are clearly immensely complicated and difficult, but they are of such general importance that we can expect them to continue to receive intensive attention in the immediate future. These investigations will need the close co-operation of chemists, physicists, and biologists, and if one were asked to put into a few words the general impression made by this comprehensive symposium, I think it would be correct to say that the papers presented and the vigorous and lengthy discussions have revealed an astonishing increase in our knowledge—chemical and biological—during the past few years. They have brought to light an equally surprising number of fundamental problems still remaining to be solved and in a remarkable way it has become clear that although at first sight some of these aspects of sucrose science may seem to be unrelated to each other, yet all the various modes of approach to the study of sucrose are beginning to fit together to give a coherent picture. The final judgement is well expressed at the conclusion of Professor Yudkin's paper when he says 'Much more work is needed, however, before we can answer all the questions that are raised by these considerations'.

I

SUCROSE AS A FOOD

Chapter One

SUGAR AS A FOOD: AN HISTORICAL SURVEY

JOHN YUDKIN

Professor of Nutritition and Dietetics, Queen Elizabeth College, University of London

Whatever may be the future of sucrose utilisation, it has been known up to the present almost entirely as a constituent of man's diet. Its deliberate production began some 2500 years ago, about one-third of the time during which man has been a producer of food. For millions of years before this, he was a hunter and gatherer of food, like all other species of non-domesticated animals alive today. His sucrose then came almost exclusively from wild fruits, and whilst it is impossible to determine how much these formed part of his diet, it is probably not far wrong to say that his consumption of sucrose could have been only a few grammes a day.

The sugar cane was cultivated in India at least as early as 325 B.C.; its cultivation then spread, slowly at first and later with increasing speed, so that by the fifteenth century it was growing in China, Arabia, the Mediterranean countries, the Canary Islands, North and West Africa, Brazil and the Caribbean. But even then it was an expensive commodity, consumed in very small amounts as a condiment by the wealthy. By the middle of the sixteenth century, its cost in England was equivalent to the cost of caviar today.

Total sucrose production is now over sixty million tons a year. About 60% of this comes from the cane, and about 40% from the beet. Sucrose from all other sources—the sugar-palm, sorghum and maple—amounts only to about 1% of the total *(Table 1)*. Thus, the average consumption of sugar is something around 15 kg a year, but there is a considerable difference between different countries. On the whole, the amount of sugar consumed by a country is an index of its average income. The wealthiest countries, such as the United Kingdom, Australia, Sweden, the United States and Canada, consume on average around 50 kg a year, whilst the poorest countries such as India, Pakistan and China consume 5 kg or less a year.

World production of sugar has grown enormously; even in the last sixty or seventy years it has increased sixfold *(Figure 1)*. Even allowing for the population growth during

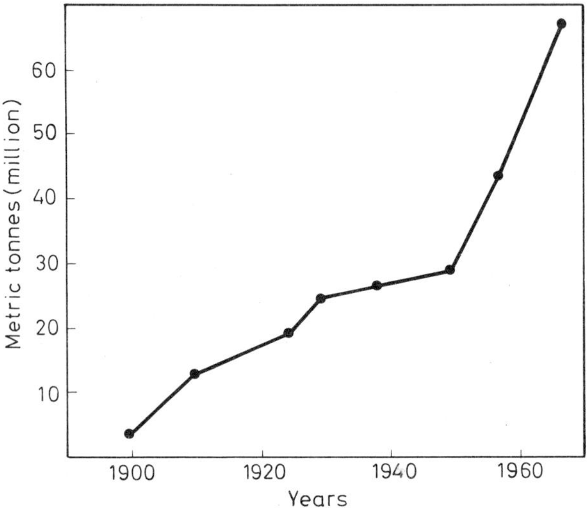

Figure 1. Increase in world sugar production, 1899–1967

that time, this is an increase of threefold in average consumption *(Table 2)*. This is a far greater increase than in any other major foodstuff—meat, milk, cereals, fats and oils *(Table 3)*. Recently the increase has been fastest in the poorest countries, for it seems that an asymptote in consumption is reached at a level of about 50 kg a year *(Figure 2)*. Consumption in the United Kingdom has increased by about 10% since 1939, but in several countries in Latin America, the Near East and Africa it has increased by 200% or 300% or more.

SUGAR AND AFFLUENCE

As with other indices of wealth, the recent rapid increase in sugar consumption in the poorer countries was preceded by a

Table 1. SOURCES OF SUCROSE

Approximate total annual production—65 million tons

Source	*Percentage of production*
Sugar cane	55
Sugar beet	44
Palm sugar	1
Sorghum	0·1
Maple	0·05

Table 2. YEARLY WORLD SUGAR CONSUMPTION

(from Viton and Pignalosa,[1] by courtesy of the F.A.O.)

Year	*Consumption*	
	kg/head	*Index*
1899	5·5	100
1909	7·5	135
1924	10·2	185
1929	12·4	225
1938	12·0	218
1949	11·8	215
1957	15·5	282

Table 3. PERCENTAGE INCREASE IN 1958 OVER PREWAR PRODUCTION OF SOME AGRICULTURAL COMMODITIES

(from Viton and Pignalosa,[1] by courtesy of the F.A.O.)

Commodity	*Percentage increase*
Cocoa	20
Cotton	20
Rice	30
Milk	35
Fats and oils	35
Meat	45
Wheat and other grains	45
Sugar	100

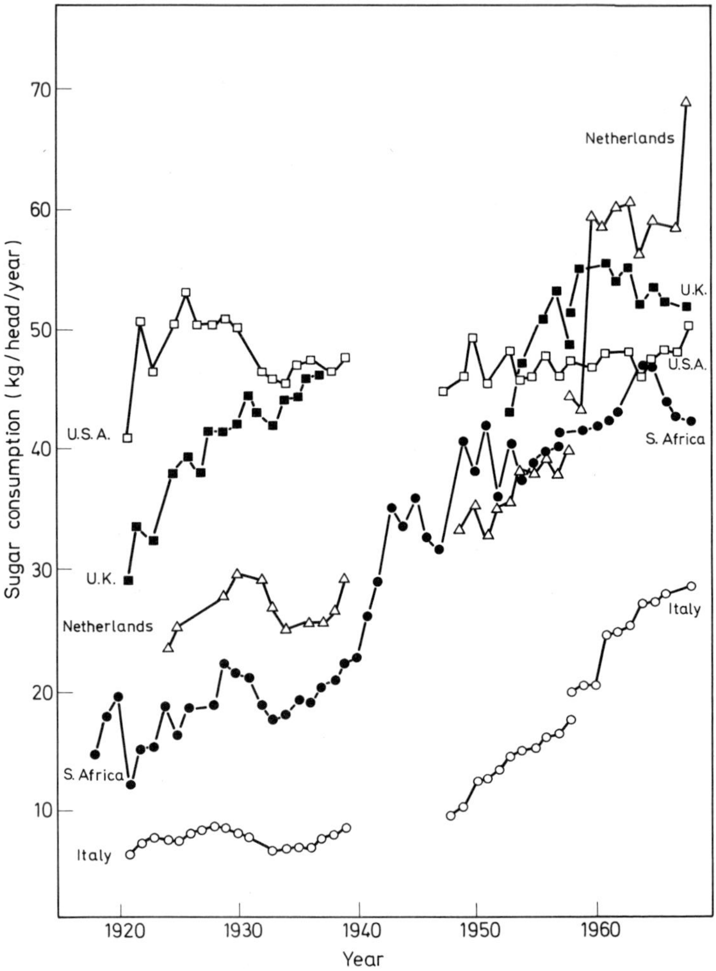

Figure 2. Average sugar consumption in five selected countries. Data from Viton and Pignalosa[1] *up to 1957 or 1958, and from the monthly Statistical Bulletin of the International Sugar Organisation from these dates (by courtesy of the F.A.O. and the International Sugar Organisation). Figures from both sources are given from 1957 or 1958, since they do not exactly coincide. A gap from about 1938 or 1939 is left in countries where there was rationing of sugar*

much slower increase, over a much longer time, in those that are today the wealthy countries. The United Kingdom provides a good example. The average annual consumption in the early eighteenth century was 2 or 3 kg; by the middle of the nineteenth century it had reached 10 or 12 kg, and now as we saw it is about 50 kg *(Figure 3)*.

Thus, there is a relationship between sugar consumption and affluence, seen both when we compare present consumption in different countries, and when we look at rising consumption in a country as it moves from poverty to affluence. The characteristics of the diets of affluence are a small rise

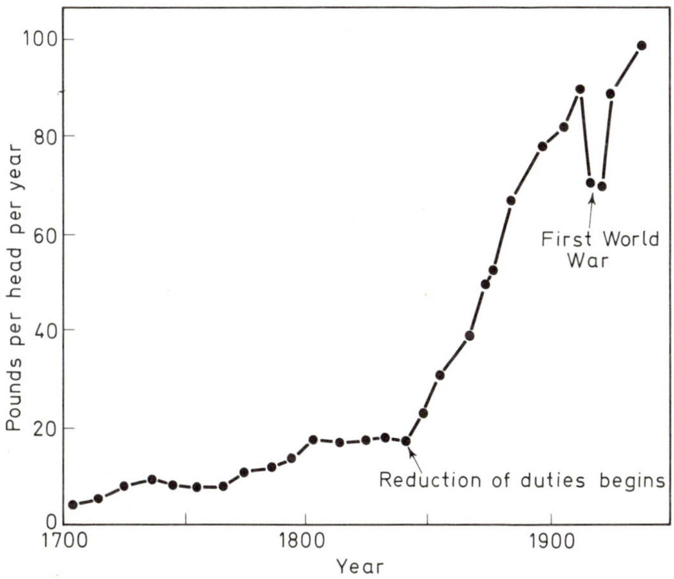

Figure 3. Annual sugar consumption in United Kingdom (from Deerr.[4] by courtesy of Chapman and Hall)

in protein, a considerable rise in fat, and a fall in starch that is about the same as the rise in sugar. Total carbohydrate is not very much different in countries of very different degrees of affluence *(Table 4)*.

Much of the sucrose in the diets of the high-consuming countries is found in manufactured foods such as confectionery, soft drinks, cakes and biscuits. In this country, about half of our sugar is manufactured, the other half is 'household sugar'; this is bought as sugar by the housewife,

Table 4. EFFECT OF AVERAGE NATIONAL INCOME ON INTAKE OF CALORIES, PROTEIN, FAT, TOTAL CARBOHYDRATE AND SUGAR[2]

Annual income (£)		No. of countries	Daily intake					
Range	Average		Calories (kcal)	Fat (g)	Protein (g)	Carbo-hydrate (g)	Sugar (g)	Sugar (% of carbo-hydrate)
Up to 50	38	6	2 060	34	52	394	35	9
51–100	82	6	2 500	61	68	437	59	13·5
101–200	148	7	2 700	73	78	456	86	19
201–350	292	7	2 970	118	86	419	102	24
More than 350	460	8	3 200	140	92	422	139	33

Values for groups of countries not weighted for population.

Table 5. SUGAR USED IN MANUFACTURE AS A PERCENTAGE OF TOTAL SUGAR CONSUMPTION

(from Viton and Pignalosa,[1] by courtesy of the F.A.O.)

Country	Year	Sugar used in manufacture
South Africa	1956–7	20
West Germany	1956–7	40
France	1956–7	42
U.K.	1958	49
U.S.A.	1957	51
Australia	1955–6	55

and used in tea and coffee, on cereals and puddings, and in cooking and baking. As with total sugar, the proportion used in manufacture has increased with increasing wealth, and is higher in the wealthier countries than in the poorer *(Table 5)*.

From these trends in sugar consumption, it seems possible to derive a few general conclusions. Firstly, the amount of sugar consumed is very much a function of wealth in relation to sugar prices. Secondly, there seems to be a limit, amounting to something like 50 kg a head a year, to the amount of sugar that is consumed in the wealthy countries, a level which most of the poorer countries are rapidly approaching. Thirdly, this high level of sugar consumption occurs at a time when the consumption of starch-containing foods has fallen; the result is that total carbohydrate intake is very little altered. Finally, the high level of sugar intake in wealthy countries is possible because a great deal is taken in sophisticated manufactured foods and drinks, in addition to that taken as sugar itself in the household.

One final point. To put dietary sucrose into its proper perspective, we should consider it in relation to the other major dietary constituents. In this country, about 50% of our total calories come from carbohydrate; in absolute terms, this amounts to about 350 g a day. Very few analyses have been made of the distribution of the different carbohydrates within this total, but there are data published by the national food survey, and we ourselves have analysed some diets in detail.[3] From this information, we estimate that the approximate carbohydrate contribution to the daily diet is 175 g starch, 140 g sucrose, 20 g lactose, and 15 g other sugars, almost entirely fructose and glucose.

REFERENCES

1. VITON, A. and PIGNALOSA, F., *F.A.O. Commod. Ser. Bull. No. 31* (1961)
2. YUDKIN, J., *Proc. Nutr. Soc.*, **23,** 149 (1964)
3. YUDKIN, J., *Am. J. clin. Nutr.*, **20,** 108 (1967)
4. DEERR, N., *The History of Sugar*, Chapman & Hall, London (1950)

Chapter Two

SOCIAL ASPECTS OF THE TASTE FOR SWEETNESS

G. W. LYNCH

Research Sociologist, Department of Nutrition, Queen Elizabeth College, University of London

If a semi-tropical origin of the human stock can be assumed, then it is likely that man emerged from regions for which his palate was suitably equipped to enjoy their varieties of fruits as well as wild honey. Man settled in colder zones presumably had the capacity for appreciating sweet-tasting food but less ability to experience it. Nature produced little in the way of sweet foods in these spheres but man's interventions came to adjust this deficiency to meet his demands.

The production of sugar was known in India perhaps a few centuries before the Christian era. About one thousand years after this epoch, the Venetians imported sugar into Europe from Syria and Egypt. The methods employed by the Egyptians in refining sugar earned them a high reputation and this probably explains why sugar-refining was not adopted in Europe until about the fifteenth century. The development of colonialism included the cultivation of sugar-cane to the extent of the world becoming dependent mainly on American sugar by the end of the eighteenth century. It was at about that time when, in addition to cane, beet-sugar as a source of supply became a practical proposition. Machinery had been developed for extracting beet-sugar and was first put into use at the beginning of the nineteenth century. By the turn of the twentieth century, both sources were making nearly equal contributions to world supplies of raw sugar. Thus, sugar that had for many centuries remained a costly luxury for the few was now within the reach of all.

SUGAR CONFECTIONERY

In addition to household consumption of sugar, other sources of supply exist in a variety of confections. Milk chocolate was invented at the end of the first quarter of the eighteenth century but large-scale production did not occur until the last quarter of the nineteenth century. The confectionery trade in general developed from small-scale domestic industry and expanded considerably only when mass production techniques and rising living standards made this possible towards the end of the nineteenth century.

Between the two world wars, the sales of confectionery and soft drinks increased enormously in the U.S.A. Doubtlessly the prohibition of alcohol between 1920 and 1932 was a factor of considerable influence in this growth.[1] More money was released that could be spent on food products and it is possible that confections were seen as substitutes for alcohol if only because both are associated with relaxation. By 1927, American consumption of confectionery had trebled in a little more than ten years. It doubled in Britain during the same period. It was during this period that convenience packs and vending machines were widely introduced to facilitate both distribution and consumption.

Today, the British eat more sweets than any other people and consume more sugar than most. Britain is now the world's largest exporter of sweets and meets one-third of world demand. In 1966, we spent £340 million on sweets and this represented an increase of 5% on the average consumption figures for 1965. This level of consumption per head is about half as much again as that for either U.S.A. or Germany. There are of course media other than sweets that meet demands for sweet-tasting food. For example, we spend £160 million a year on cakes, and this market grows at the rate of 7% per annum. However, against these increases in consumption levels, household sugar consumption has contracted since 1958 and in 1965 it was back to the 1956 level. Further, the consumption of jams and jellies has declined over the last decade by an annual rate of 2%.

NATIONAL VARIATIONS IN SUGAR CONSUMPTION

It is difficult to determine why so much confectionery and sugar is consumed in Britain when compared with other western countries. Nevertheless, certain factors may help in part to explain this phenomenon. For example, perhaps Americans are more concerned than the British about the relationship of food to health and appearance. The collection of accurate data on health and slimming foods is fraught with endless problems.[2] To interpret comparisons between groups in which the function of social variables may differ makes for further complication. However, one partial indication that gives support to the suggestion relates to the growth of the 'Weight Watchers' membership in the two countries. An estimate based on rate of membership shows that when compared with Britain, it is about fifteen times greater in the U.S.A.[3] As far as this may be an index of more interest shown about the need to avoid excessive calorie intake, it is assumed that confectionery will be among the foods to be resisted.

Fruit consumption is a distinguishing factor between Britain and other countries in Western Europe. With the exception of Eire, the British consume less fruit per head per day than any other country.[4] In many parts of Europe, varieties of fruits are available at cheaper prices than one would expect to pay here, whereas sweets and cakes tend to be cheaper in Britain. For example, in the event of our joining the E.E.C., the cost of sugar would increase by 66% and confectionery would reflect this increase in cost. Meanwhile, prices have been reduced in many stores on the withdrawal of resale price maintenance.

What is known as 'plain English cooking' may make the palatability of sweet-tasting food a necessary relief in our daily food patterns. One survey[5] showed that 92% of housewives usually cooked plain English food which was said to possess such qualities as simplicity, cheapness and cleanliness. It was thought also to avoid any suggestion of greed or self-indulgence. This type of cooking was overwhelmingly associated with main dish meals which are functional and tradition-bound. The report points out that importance is given to pleasurable aspects of food—especially pleasure to the palate—when housewives become involved in producing puddings.

SUGAR IN THE MODERN SOCIETY

Industrialised societies of today are increasingly involved in demands for full employment, affluence and leisure. These will impinge on eating habits. As more women in western society go out to work and husbands and children have midday meals at work or school, housewives will tend to opt out of kitchen chores in the evening. If they satisfy themselves that the serious main meal has been eaten elsewhere, a light simple evening meal can border on the frivolous. Today's widespread relaxation in the home or the pursuit of social activities outside of it invite the association of pleasure and convenience with the evening meal. This could be particularly relevant to sweet courses which can so easily accommodate both factors of association. Further, this may not apply to the British alone. Moulin's[6] examination of the European's changing gastronomy suggests that rituals like home-made jam-making and indeed traditional family cooking in general are gradually going the way of knitting and sewing. In fact, he claims that tomorrow's industrial society will exclude the requirements of everyday family cooking and that gastronomy will become the instrument of conviviality and leisure.

How far levels of confectionery consumption are due to palatability and how far they arise from persuasion and conditioning by the industry is a matter for conjecture. The taste of sweetness appears to be universally considered as a source of undiluted pleasure to the palate. Sweetness has come to be connected with emotive value judgements of what are thought to be psychologically, socially and materially desirable qualities and ends. Terms like 'sweetie', 'a sweet person', 'sugar-daddy' and 'a piece of cake' suggest how powerfully sweetness connotes desirability. It is a taste that is highly associated with affection and reward and few would doubt the enjoyment afforded to adults related to children and men to women when proffering and receiving gifts of confectionery. It seems that for many there is an analogy between the relationship of sweet-tasting foods to overall food intake and leisure as it relates to overall cultural activities and responsibilities required of the individual. Both sweet-tasting foods and leisure are socially permissible forms of relief from necessity and discipline.

CONFECTIONERY AND ADVERTISING

Confectionery manufacturers are naturally interested in encouraging the maximum consumption of their products. Chocolates loose in boxes and unit divisions in a block of chocolate are made in such sizes as to be equal to a small bite and, together with a variety of different flavours, physiological fatigue is thus avoided and consumption is speeded up. Sucking a sweet is more fatiguing than eating chocolate and so sweets are usually smaller and lighter in comparison.[7] Expenditure on television advertisements for sweets and chocolates in Britain is the highest for any single product and in 1967 was 5% greater than expenditure for detergents.

The medical profession is usually of one voice in condemning the effects of excessive sugar and confectionery on general and dental health. I have already given some indication of what health educationists must contend with in endeavouring to bring about a conformity with medical advice. An increasing range in the variety of confections available coupled to modern methods of communication will the more effectively influence those who are vulnerable to the persuasiveness of the manufacturer. The statistics given of production and consumption in Britain cannot be unrelated to the vast expenditure on advertising to which reference has been made.

Disraeli described sugar as that which 'charms infancy and soothes old age' and in this, he accurately, if unwittingly, pointed to the most profound problem related to hazards of over-indulgence in the consumption of sweet-tasting foods. 'Addiction' is likely to be formed in the most vulnerable of all social groups, the young. Those who are responsible for the continuance of linking confections with affection and reward are amply encouraged in their gratifications and indulgences. Much has to be done to counteract the effects of these influences upon those in their formative years. Patterns and associations developed during this time are inclined to be durable whether the habits are good or bad.

REFERENCES

1. DUNN, R. W., *The Confectionery Industry: Encyclopaedia of the Social Sciences,* The MacMillan Co., New York, **I**, 310 (1949)
2. DEUTSCH, R. M., *The Nuts Among the Berries,* Ballantine Books, New York, 273–274 (1968)
3. *Financial Times,* 17 April 1967 and 5 April 1969
4. *Food Balance Sheets, 1960–1962,* F.A.O., Rome (1966)
5. *Food and Cooking in England:* A study based on research undertaken by the British Market Research Bureau, J. Walter Thompson Co. Ltd. (1964)

6. MOULIN, L., 'The European and his Dinner Table', *European Community*, **1**, 8 (January 1967)
7. RENNER, H. D., *The Origin of Food Habits*, Faber and Faber, London, 118–122 (1944)

Chapter Three

SUGAR AND FOOD CHOICE

R. H. J. WATSON

Research Psychologist, Department of Nutrition, Queen Elizabeth College, University of London

Several of the papers in this symposium will put forward views to the effect that dietary sucrose is a mixed blessing. On the one hand it is claimed that attractive and appetising foods may be prepared with its use, whilst on the other hand it is claimed to leave in its trail pathological conditions of various kinds—dental caries, coronary thrombosis, etc. We know that individuals differ in the amounts of sucrose that they take in their diets, and it becomes therefore more than of mere academic interest to try to discover the factors that determine the level of sucrose intake of an individual. Since in our own society the individual is to some extent free to choose what he will eat, at any rate when he is adult, one of the ways into this problem is to investigate the relationship between sugar and food choice.

In discussing this relationship it must be remembered that there are two quite distinct types of problem involved. The proportion of sugar in the diet may be merely a reflection of the amount of carbohydrate in the diet, the form the carbohydrate takes being largely immaterial. On the other hand the amount of sugar present may reflect a preference for this form of carbohydrate over other forms, e.g. starch. Frequently this distinction is not always borne in mind and confusion occurs in the literature. For example, in a recent paper[1] investigating differences in 'craving for sugar' in different strains of rats, the experimenters made no attempt to use other forms of carbohydrate. In other words what they were studying was the amount of carbohydrate selected in the total diet, and all they were really entitled to talk about would be 'carbohydrate proportion' and not 'craving for sugar'. Let us now consider these two types of problem separately.

PROPORTION OF CARBOHYDRATE IN THE DIET

It is well known that the composition of diets selected by individuals may differ very widely between one individual and another with respect to the proportions of the major nutrients present. A great many factors—economic and social as well as past dietary experience—may help to determine these differences.[2] From a nutritional point of view, provided certain minimum requirements are met, e.g. amount of protein available for growth and maintenance, there is considerable flexibility in the type of nutrients which may be used to provide the energy component of the diet. Unfortunately, owing to the social and psychological influences operating, it is very difficult to answer the question 'To what extent are there factors affecting the proportion of carbohydrate selected in the diet which reflect individual variation in biochemistry and physiology?'

Again this question is not one of mere academic interest. Let us suppose that one is trying to encourage a patient to reduce carbohydrates in his diet because he is beginning to show signs of obesity. It is very pertinent to know whether the liking for carbohydrate is determined largely by social and psychological factors, or whether it is determined by biochemical and physiological mechanisms which in his case predispose him to prefer a diet rich in carbohydrate.

However, it is possible to show that such effects do operate in experimental animals where, of course, the animal can be under the experimenter's control from conception. For example, changes in hormone activities either naturally occurring as in pregnancy and lactation[3] or experimentally produced as in adrenalectomy[4] have been shown to affect the proportion of carbohydrate selected in the diet.

It has been shown by the Prague group[5, 6] that individual differences in the relative proportions of carbohydrate and protein selected in the diet are related to behavioural characteristics. Rats selecting a relatively high protein diet showed less spontaneous activity and less exploratory behaviour, and developed conditioned responses more slowly, than did rats that selected a relatively high carbohydrate diet. In terms of Pavlovian neurophysiology the high-carbohydrate eaters would be characterised by an 'excitatory' nervous system in contrast to the 'inhibitory' one exhibited by the high-protein eaters. They have also shown that the reverse process is true, namely, that feeding a particular type of diet may produce changes in the 'excitatory'/'inhibitory' level of the animal.

PREFERENCE FOR DIFFERENT FORMS OF CARBOHYDRATE

Although there are very obvious differences in the subjective experiences associated with the tastes of different forms of carbohydrate, e.g. sucrose and starch, it used to be thought that from a nutritional point of view they could be regarded as equivalent in that eventually they all ended up as glucose. However, in recent years, as will become obvious during this symposium, it has been demonstrated without any doubt that this view is no longer tenable.

This again raises the interesting point of whether differences in preference for different forms of carbohydrate might reflect variations in biochemistry and physiology between one individual and another. Again in the human field there are problems of social and psychological influences, particularly when sucrose containing foods are used in social situations to accompany pleasurable activities or social approval. Nevertheless, the need to know more about 'the sugar eaters' and their biochemical, physiological and psychological characteristics is of real practical value, in view of the dangers of sucrose consumption to the health of an individual. If, for example, our hypothetical patient had signs of ischaemic heart disease and we wished to encourage a reduction in sucrose intake, it would be pertinent to know whether we were dealing with a liking which had resulted from past social and psychological pressures, or with a liking which was reflecting some possibly innate biochemical requirement.

Before plunging into the human field we thought that an investigation of the factors related to carbohydrate preference in experimental animals would be valuable in suggesting aspects worth pursuing in human studies. A number of studies have been undertaken which have suggested several factors worth further consideration.

Using 100 day-old rats we have demonstrated[7] that male rats, of the Lister Hooded Strain maintained at this College, when presented with a choice of diets varying only in the type of carbohydrate present, select a diet which contains relatively more dextrin (dextrinised corn starch) and less sucrose than the diet selected by females. This suggests that hormone differences may influence the carbohydrate preference. This finding has been confirmed in subsequent work[8] in which the difference was shown to develop during the 'adolescent' period. A further confirmation came from the fact that even within a group composed of animals of only one sex, there were small but positive correlations between preference for dextrin over sucrose, and other 'masculine' characteristics such as greater growth rate and less exploratory behaviour.

RELEVANCE TO MAN

At this point one should perhaps ask the relevance to the human field. One hears a number of suggestions made of there being sex differences in food preferences, one such being that

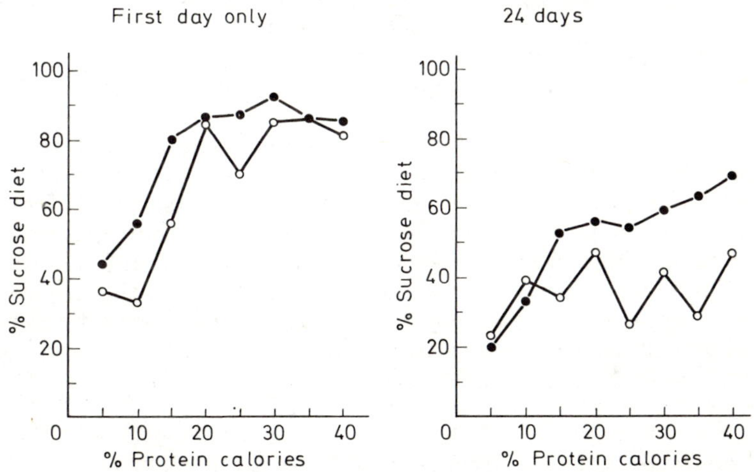

Figure 1. Effect on carbohydrate preference of varying protein: ● against carbohydrate. O against fat

women prefer more savoury foods and men take more sugar. We have certainly found in some studies to be reported later that men do add more sugar to their drinks than women, but one must ask how far this is due to the constant pressures from health and fashion propaganda of one kind and another to stay slim by cutting down on sweet things, rather than a reflection of differences in biochemical requirements.

We next looked at factors likely to affect this preference. Studies have shown that when protein is increased at the expense of carbohydrate, rats prefer a higher proportion of sucrose to starch as the carbohydrate.[9] On the other hand, when protein is increased at the expense of fat, this relationship is shown mainly on the first day and is thereafter much reduced *(Figure 1)*.

In both types of experiment there is on the first day a greater preference for sucrose over the entire range of protein levels than is shown subsequently.

What implications are there here for future human work? Clearly discussions of dietary sucrose levels must always be made in conjunction with a discussion of other dietary components. We do not as yet know why in the case of the rat this increase in sucrose preference as protein level is increased (or carbohydrate level reduced) should occur. It may be a function of the fact that, as the protein changes, the taste-texture sensations may change in such a way that sucrose makes the diet more acceptable. For example, in these experiments a finely powdered casein was used as the protein source. In large quantities in the diet the taste-texture may be such that the addition of 'sweet granularity' may overcome any adverse aspects of the high casein content. In the human field there is of course the equivalent example of adding sucrose to overcome the tartness of acid fruits, which fools many unsuspecting sucrose-watchers into thinking that 'bitter-lemon' drinks are acceptable for them as they cannot presumably contain sucrose. The second point of the greater preference for sucrose on the first day compared with subsequent days was confirmed in the next experiments and its implications will be discussed there.

'TOGETHERNESS' IN FOOD PREFERENCE

A more interesting factor was found to be the conditions of husbandry under which the animals were reared.[8]

Earlier work had demonstrated the phenomenon of 'social facilitation' of eating in rats.[10] That is, rats fed in isolation from each other eat less than when fed in groups: a phenomenon not unknown in human society. What we were interested in was whether such social factors could influence carbohydrate preference as distinct from total amount eaten. Animals housed singly preferred more sucrose to dextrin as the carbohydrate than did animals housed in pairs. *Figure 2* shows the results for male pairs but this was true for either sex and for mixed pairs. In all instances, on the first day the

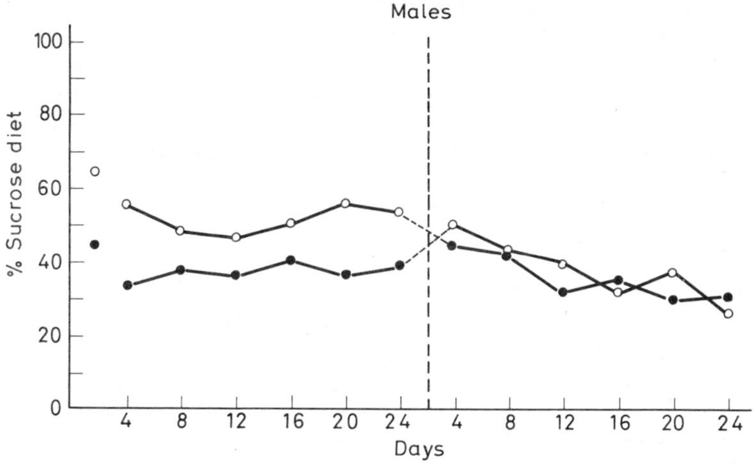

Figure 2. Effect on carbohydrate preference of varying husbandry: ● 2 per cage. ○ 1 per cage

preferences shown were shifted towards relatively more sucrose. After 24 days the husbandry of the animals was reversed —those which had been in pairs were now separated and those which had been housed singly were now put in pairs. The preferences now shifted towards that which had been previously demonstrated by the other group of animals. The results suggest that the differences might be the result of 'stress' if one takes the view that moving animals into new cages, changing diets, etc. constitute 'stress'[11], and that for rats which are group living animals the isolated state is more 'stressful' than the group state.

In human terms it has been suggested that sweetness in food provides a pleasant taste and that it can be used to allay anxieties and to increase the feeling of well-being, hence the custom of giving sweet things as presents. This has not yet been shown experimentally.

Another of the factors studied has been the effect of early dietary experience of particular carbohydrates.[12] There is evidence that such experience does affect subsequent preferences. The data suggested that there may be a 'critical period' during which experience of particular carbohydrates may produce marked effects of subsequent preferences. This was further confirmed in subsequent experiments in which those animals which had had access to diets prior to weaning showed significantly different subsequent preferences from those which had been denied access until after weaning. The existence of 'critical periods' in the development of human infants is well established. One is led to ask whether or not there are 'critical periods' in terms of dietary experience in the life of the baby which might have important effects on its preferences in later life. This possibility is of obvious importance to manufacturers of baby foods, some of which require the addition of sucrose.

RELEVANCE TO ANIMAL FEEDING PRACTICE

Leaving the human field for a moment, there are several problems which have to be faced by the makers of commercial animal feeds and by experimentalists who need to prepare highly purified diets of one kind or another. Since there is a wide range of individual differences in food selection, one is led to ask the question 'Is there an optimum mixture of different carbohydrates which provides the most acceptable diet to most animals?'. This would suggest that the consumption of such a diet would be maximal, and that a less optimal combination would lead to decreasing acceptability and hence result in lower food intake. However, studies have shown that it is better to provide the animal with a ready-mixed carbohydrate fraction of the diet, than to allow the animal to select its own mixture from diets containing the carbohydrates separately. Although as yet we have only investigated a few points on the continuum, something of the nature of a mixture of 50% starch and 50% sucrose results in increased food intake and growth compared to allowing the animals to select from diets containing these separately.

We have made a small beginning in the human field. It is possible to assess by questionnaire the amount of sugar added by people to their beverages each day.[13] This is a conscious intake of sucrose, as distinct from the hidden sucrose which may be present in other foods and drinks. We have discovered a small but significant correlation between beverage sucrose intake and 'extraversion' as measured by the Eysenck Personality Inventory, in two groups of male students. The absence of such a relationship in comparable female students may well be due to the fact that restriction of sucrose occasioned by dieting for fashion purposes was prevalent among them. At least we know that relationships between dietary and psychological variables are not out of the question, and in the sugar context could be of immense help in knowing the type of factors to be dealt with in any attempt to modify a person's sucrose intake.

No doubt other aspects of personality will be shown to be more closely related to sucrose preference or to total proportion of the diet as carbohydrate. In particular the special personality types of Rosenman and Friedman[14] which they have shown to be related to liability to coronary thrombosis are worth investigating.

REFERENCES

1. VARTIANEN, I., MÄKELÄINEN, T., NIKKOLA, E. and NIEMININ, E., *Annls Med. intern. Fenn.*, **56,** 137 (1967)
2. YUDKIN, J., *Lancet,* **1,** 645 (1956)
3. RICHTER, C. P., *Harvey Lect.,* **38,** 63 (1942)
4. RICHTER, C. P., *Proc. Soc. exp. Biol. Med.,* **48,** 577 (1941)
5. LÁT, J., *Proc. Nutr. Soc.*, **18,** xxiv (1959)
6. FRANKOVÁ, S., *Čslká Psychol.*, **10,** 111 (1966)
7. WATSON, R. H. J., *Proc. Nutr. Soc.,* **23,** xli (1964)
8. WATSON, R. H. J., *Proc. Nutr. Soc.*, **25,** xiii (1966)
9. SHIM, K. F., 'The Relationship between the Composition of the Diet and Carbohydrate Preference in the Rat', Ph.D. Thesis, University of London (1968)
10. HARLOW, H. F., *J. genet. Psychol.,* **41,** 211 (1932)
11. STEINBERG, H. and WATSON, R. H. J., *Nature, Lond.,* **185,** 615 (1960)
12. GREENFIELD, H., *Proc. Nutr. Soc.,* **28,** 19A (1969)
13. YUDKIN, J. and RODDY, J., *Br. J. Nutr.,* **20,** 807 (1966)
14. ROSENMAN, R. H., FRIEDMAN, M., STRAUSS, R., WURM, M., JENKINS, C. D. and MESSINGER, H. B., *J. Am. Med. Ass.,* **195,** 86 (1966)

Chapter Four

SUCROSE AND THE FOOD MANUFACTURER

MAURICE BROOK

Chief Research Manager, Basic Research Laboratories, Beecham Products (U.K.)

Sucrose is one of the purest food raw materials that is available in large quantities to the food manufacturer, the housewife and the caterer and one of the most consistent in composition.

Food manufacturers use sucrose because it helps them make delicious foods that people will buy because they enjoy eating them. Most human beings like sweet things but this characteristic is not universal in the animal kingdom. For example, cats, some birds and some species of fish do not respond to sugar sweetness, and the rat appears to have a high preference for maltose, a sugar that gives man little sensation of sweetness.[1] For man, the quality of sweetness imparted by sucrose is unique in its intensity, its delectability and its cleanness. Unlike non-caloric sweeteners, sucrose does not induce an unpleasant taste reaction at any concentration, but it imparts a clean, sweet, taste over its entire range of concentration from near zero to 100%.

Just as the housewife is provided with a variety of types of sugar—granulated, cubed, castor, etc.—so the sugar refiner supplies the food manufacturer with a variety of sucrose products tailored for particular uses *(Table 1)*. For example, sugar for canned foods is specially selected for its low count of bacteria that may cause spoilage. Mineral water sugar is a granulated sugar specially refined to have an extremely low ash content and a good colour, and a low tendency to foam or to discolour when heated. It is especially suitable for high quality confectionery such as clear mints, fruit preserves and pharmaceuticals. About 65% of the sugar used by manufacturers is supplied in bulk in granulated form, but the basic grades of dry sugar are also supplied in liquid form. Mixtures of sucrose and glucose syrup or sucrose and invert sugars are also available and widely used by food manufacturers.

During 1965/66, at least 2 648 000 tons of refined sucrose was used in food in the U.K. Some of the sugar used by the

Table 1. SOME BASIC GRADES OF INDUSTRIAL SUGARS

Description	*Characteristics*	*Some Uses*
Mineral Water	Pure white, low ash	Preserves, drinks, high quality foods, medicines
Standard Granulated	High purity, quality, consistency	Manufacturing
Canners	Low bacterial count	Canning
Manufacturing	(*a*) white, off-white, yellow (*b*) brown—various crystal sizes	Baking, confectionery
Liquid sugars	Equivalent grades to above	as above
also Sucrose/Glucose/Invert mixtures		

Table 2. REFINED SUGAR IN U.K. FOOD
Mean of 1965 *and* 1966

	Long tons
Disposals	2 648 000
Less content of exports	75 000
	2 573 000
Plus content of imports	95 000
TOTAL	2 668 000

food manufacturers leaves the country in exported food products *(Table 2)*. From the declarations of manufacturers and trade associations it can be calculated that rather less than half (42%) was used by food manufacturers and more than half (58%) by housewives and caterers *(Table 3)*. Our total consumption of sucrose, at nearly 1 cwt/head/year, is high compared with most other countries *(Table 4)*. About two thirds of our sucrose is derived from cane-sugar and the remaining one third from beet-sugar, mainly home grown.

Sucrose is not the only material available to the food manufacturer as a source of sweetness. Although it is sweeter than most other carbohydrates, except fructose, it is much less sweet than saccharin by a factor of 300 to 700 and than cyclamate by a factor of 30 to 60 or in some cases 120[2], *(Table 5)*. However, unlike the non-caloric sweeteners, sucrose has a remarkable ability to enhance flavour in food

Table 3. DISPOSAL OF REFINED SUGAR FOR FOOD IN THE U.K.

	Mean of 1965 *and* 1966 Thousand long tons	%	%
Chocolate and sugar confectionery	359	32·5	
Cakes and biscuits	175	16·0	
Jams and marmalades	137	12·5	
Soft drinks	130	11·5	
Beer and British wines	80	7	
Syrup and treacle	53	5	
Canned vegetables, fruit and soups	44	4	
Table jellies	33	3	
Condensed milk	30	2·5	
Ice-cream	18	1·5	
Cereal breakfast foods	15	1·5	
Cider	12	1·0	
Pickles and sauces	14	1·0	
Cake and bun mixtures	5	·5	
Canned puddings	5	·5	
Quick frozen and vegetables	1		
	1 111*	100	42
Household and catering (by difference)	1 537		58
TOTAL DISPOSAL	2 648		100

*about 75 000 tons exported, 90–100 000 tons imported

products even when used in small amounts. In preserved meats, for example, the concentration present is usually less than 1% and this is too low perceptibly to increase sweetness but it tempers the salt note in such products. The flavour-enhancing capacity of sucrose may be partially related to its ability to promote the dissociation of weakly ionised compounds.[3]

One can match the sweetness of 1 lb of sugar with about 1 g of saccharin or 10 g of cyclamate. This dramatically illustrates one other characteristic besides sweetness which contributes to the importance of sucrose in food manufacture, for sucrose provides bulk as well as sweetness. Indeed, it possesses many characteristics besides sweetness and some of these are summarised in *Table 6.* The relative importance of these characteristics varies with the user industry.

Just over one million tons of sucrose, of all types and grades, is used in the production of manufactured foods in the United

Table 4. RANGE OF AVERAGE SUGAR CONSUMPTION 1967

kg/head

World Average	18·9	Oceania	54·1
		North America	48·2
U.K.	52·2	Europe	38·8
		Central America	36·6
		South America	31·5
		Africa	10·8
		Asia	6·8

Table 5. RELATIVE SWEETNESS

Sucrose = 1

Fructose	1·1
Glucose (dextrose)	0·61
Sorbitol	0·54
Liquid Glucose (BPC 1963)	0·23
Lactose	0·16
Cyclamate	30–60–120
Saccharin	300–700

Kingdom and about 7% of this sugar is exported in finished products. A major portion, nearly one third, is used in the manufacture of chocolate and sugar confectionery, followed by cakes and biscuits (16%), then, around the 10% level, jams and preserves and soft drinks, decreasing to brewing (7%), treacle (5%), canned vegetables, fruits and soups (4%) and so on *(Table 3).* In the United States, confectionery use of sucrose is less dominant than in the United Kingdom and is about equalled by the use of sugar in soft drinks.

SUCROSE IN CONFECTIONERY

Chocolate and sugar confectionery manufacture, accounting for nearly one third of sucrose used in manufactured foods, is linked with our performance as the world's champion eaters—mainly, I believe, because in this country it is of high quality and gives good value for money. In 1967 we consumed an average 7·7 oz/head/week (220 g) of chocolate and sugar confectionery, to achieve a total annual consumption of over 614 000 tons, for the manufacture of which 355 280 tons of sucrose was used.[4] A study in 1964 by McHugh, McEwen and

Table 6. SOME CHARACTERISTICS OF SUCROSE IN FOOD

Increases—Sweetness	*Enhances*—Flavour
Osmotic Pressure	Appearance by improved clarity, lustre, gloss
Viscosity	
Boiling point	
Moisture retention	
And—Provides calories	Provides bulk, 'body'
Affects the solubility of other ingredients	
Imparts plasticity	Assists emulsification
Can be fermented	Assists colour development
Can be crystallised and the crystal size controlled	
Penetrates other ingredients—e.g. fruit and vegetables	
Depresses freezing point—hence ice-cream	

Hitchin[5] reported that the mean stated consumption of sweets by over 2500 13-year-old children in Dundee ranged from 6 oz minimum to 35 oz maximum/head/week, with an average of $17\frac{1}{2}$ oz.

It is a tribute to the economic efficiency of this section of our industry that the increase in manufacturers' prices during the past eight years has been only a fraction over 2% per annum, whereas the imposition and subsequent increases of purchase tax have increased the cost to the consumer by another 22%. Further evidence of the economic efficiency of this section of the industry is provided by its export performance. It supplies about 30% of the world market and in 1968 earned over £30 million from exports, an increase of 22·8% over the previous year, itself a record.

It is the bulk provided by sucrose and its ability to exist in different amorphous or crystalline states which, in addition to sweetness and ready solubility in water, provide the foundations for its largest manufacturing use in chocolate and sugar confectionery. A saturated sucrose solution contains

66·4 parts of sucrose at 20°C and 76·4 parts at 70°C; supersaturated solutions are easily formed.

A smooth, clear, hard, boiled sweet is around 96–98% carbohydrate solids, mainly sucrose, in supersaturated solution in water, cooled to make a 'glass'. By varying the strength of the sugar solution, different finished products with markedly different textures can be produced and the sugar boiler's thermometer is a descriptive guide to the influence of temperature on sugar solution characteristics which is so skilfully exploited by the sugar confectioner.

Pure sucrose solutions rapidly become acid on heating and some of the sucrose is 'inverted' to fructose and glucose. The process can be accelerated by adding acid or fructose or invert sugar. The degree to which this process continues affects the characteristics of the finished sweet. As inversion of the sugar proceeds the finished product becomes less brittle and more hygroscopic, a new flavour note is introduced and the colour darkens through deepening shades of straw to dark brown. Sugars and amino acids yield dark, odorous and richly flavoured products when heated together, for example in making caramels or malted cereals.

Plain, hard, boiled candies consist entirely of sugar usually with glucose syrup, flavours and some acid. The solution is cooked to over 300°F, or some 30°F lower under reduced pressure, so that the final moisture content is around 1·5% w/w. The minor constituents of the sugar used can markedly affect the properties of the finished product. A clear mint derives its diamond like brilliance and cut glass clarity from meticulous attention to detail in selection of the raw materials. For example, the sucrose used must have an extremely low ash content, and the recipe be prepared with soft water.

As a concentrated solution of sugars is allowed to cool it becomes supersaturated and thus increasingly unstable until it is completely solid. This temporary instability is used to advantage by the confectioner who, by aerating the cooling syrup, can induce crystallisation of the sugars to produce an opaque fondant suitable for a peppermint cream or a chocolate cream centre. By using invertase, enzymic conversion of sucrose can be achieved inside a fondant centre covered in chocolate to produce a liquid centre chocolate cream. By adding fat and/or milk further to the formula, texture and flavour changes can be introduced. All these ways further contribute variety to the many thousands of different kinds of sweets available to us and the success of these measures depends largely on the presence of sucrose as a major component of these products.

CAKES AND BISCUITS

Cakes and biscuits contributed £11 million to export sales during 1968, and in the manufacture of cakes and biscuits sucrose not only provides sweetness and bulk but also adopts a new role as a substrate for the rapid production of carbon dioxide during yeast fermentation. Sugar and yeast with free moisture constitute an active raising agent for bread dough. Sucrose forms a complex with gluten in flour which enables it to expand easily to produce a light texture. Crystalline sucrose creamed with fats greatly assists air dispersal and contributes in this way to desirable texture development. Sucrose also plays a part in the development of a good crust colour on bread and pies.

Protein coagulation is delayed when sugar is present and this delay makes it possible to achieve a smoother texture in the finished product, such as custard. The delicious and unusual texture of a meringue owes its properties in part to the ability of sucrose to strengthen the protein framework and in part to its assisting in the inclusion and rentention of a very high proportion of air in the whipped egg white mixture.

A not unimportant reason for the safety to health of the products of the food manufacturing industry is the high purity of the grades of commercially available sucrose, which is supplied free from pathogenic bacteria and can be supplied free from spoilage organisms.

SUCROSE AS A PRESERVATIVE

Sucrose is, of course, a valuable preservative in its own right. Through osmotic activity, it effectively lowers the moisture available to support microbial growth. The commercial preservation of milk with only mild heat treatment by adding sugar and concentrating to 70% soluble solids content as condensed milk is a relatively recent development in manufactured food by contrast with, for example, the butter and cheese industries. An English patent was granted in 1810 to de Heine for a method of heating milk in an open vessel and preserving it by the addition of sugar, and other patents were granted between then and 1850; the commercially successful manufacture of sweetened condensed milk under the Eagle brand label was begun in 1856. The development of condensed milk is a reminder that our use of sucrose is relatively brief in historical time and the growth of much of the modern food manufacturing industry is linked to the availability of sucrose at economic prices. In 1319 sugar was £20 per lb. Today its price is less than 1% of that.

Condensed milk accounts for only 2·5% of the sucrose used in the U.K. food manufacturing industry *(Table 3)*. The Condensed Milk Regulations, 1923[6], insist that sweetened, condensed milk shall indicate on the label that sugar has been added, and minimum percentages of milk fat and milk solids are also specified for this product.

When sucrose is being employed as an osmotic preservative, the substrate containing sucrose in solution is usually concentrated to above 70% soluble solids content. If the product is acid, say below pH 4·5, then concentration to about 65% soluble solids is sufficient. The manufacture and distribution of jams and preserves depends on this high sugar, high acid principle of preservation. The flavour of fruits can be faithfully, and deliciously, retained by this method, especially with concentration under reduced pressure which avoids destruction of flavour and aroma associated with high boiling temperatures. Sucrose-preserved fruit has brilliance which attracts the eye, flavour which satisfies the palate, and an easy spreading texture free from grittiness because the sucrose remains in solution. Some raw sugars and syrups which have been heated have a low capacity to support microbial activity, and this reinforces the osmotic preservative effect. The composition of jams and preserves is controlled by the Food Standards (Preserves) Order 1953[7] which specifies minimum fruit contents for the different types of jams, and a minimum solids content of 68½%, or 65% if the product is packed in hermetically sealed containers.

Sugar-preserved products such as jams and marmalades have a good shelf life and continue to keep in good condition during the normal period of use after the package has been opened. It is the local lowering of soluble solids content, through condensation for example, that is usually responsible for mould growth on the surface of jams. Incidents of this kind are more common with home-made products than with those made by the food manufacturer. I suspect that a good many home-made preserves would fail to satisfy the Regulations which the food manufacturer must observe.

Yeasts are ubiquitous food spoilage organisms and some of them will tolerate, or become adapted to tolerate, relatively high concentrations of sucrose. The spoilage of sweetened condensed milk is usually due to sucrose-fermenting yeasts. *Torula lactis condensi* will vigorously ferment and produce burst can ends[8], whereas *Torula globosa* has been identified as a cause of distention of the can ends rather than bursting.[9] Sucrose is not the most effective sugar to use for microbial inhibition. Fructose and dextrose are almost twice as effective as might be expected from a consideration of their relative molecular weights, 180 compared with 342 for sucrose[10], but the high solubility of sucrose usually gives it the advantage in practice, except where relatively low concentrations of sugar are usual—as in soft drinks for example.

The composition of soft drinks in this country is determined by the Soft Drinks Regulations, 1964[11], which, among other things, specifies a minimum quantity of sugar which must be added to most soft drinks. Sugar as defined in the Regulations can be any carbohydrate sweetener. In practice, at present, sucrose is almost universally used. Excepting low calorie drinks, a ready-to-drink product must contain, by law, at least $4\frac{1}{2}$ lb of sugar in 10 gallons; a product for consumption after dilution must contain at least $22\frac{1}{2}$ lb sugar in 10 gallons. For semi-sweet soft drinks, the values just quoted are halved.

Sucrose in soft drinks provides sweetness and mouth-feel, enhances flavour and provides food energy. It also creates some hazard because fermentation is an ever present risk not entirely eliminated by the inclusion of permitted preservatives in the finished products. The storage tanks of syrup used in manufacture represent another site where microbiological vigilance is essential. Fortunately, the sugar-supplying industry helps the manufacturer by providing special grades of sugar for particular uses, and sterile sugar syrups can be provided for use in soft drink manufacture, just as a canner's grade of sugar that is free from spoilage bacteria is available.

I will not dwell on the other minority uses of sucrose. Brewing is a special example where fermentation of the right kind is the core of the industry and sucrose is used as a supplementary energy source for the fermentive yeasts. In the other uses, sucrose provides sweetness and some bulk. In ice-cream it contributes to product quality by its ability to depress the freezing point. Under a 1953 Order, ice-cream had to contain not less than 10% sugar.[12] This requirement was replaced in 1959 by one which prohibited the use of any artificial sweetener in ice-cream.[13]

In table jellies the osmotic protection afforded by the presence of sucrose normally ensures good shelf-life, but some of you may have heard Mr. H. J. Bunker's delightful story of his encounter with osmophilic mould on a table jelly with a sucrose content of over 60%. This pest could be cultured only on media of high sugar content and was eventually identified as *Xeromyces bisporus,* a mould isolated only once before—from a stick of liquorice in New South Wales![14]

SUCROSE SUBSTITUTES IN FOODS

In all foods, sucrose has a remarkable capacity to enhance flavour in a delicious way, particularly in cereal products, fruits, coffee and chocolate. Moreover, since it provides bulk relatively cheaply, it has been an essential ingredient in many manufactured foods since food processing on an industrial scale first began. The non-caloric sweeteners cannot replace it in many types of food.[2] On average, sucrose from all sources now provides 17% of our daily calorie intake compared with 15% pre-war, an increase of over 10%.[15]

We would probably not be meeting here today if respected observers had not voiced doubts about the wisdom of us consuming so much sucrose. Should these doubts about consumption be confirmed, the food manufacturer has a formidable task to reduce his manufacturing use of sucrose. I will therefore look at some of the evidence as we see it and speculate on remedies. Two main areas of suspicion exist; first, that sucrose is an especially potent agent in the aetiology of dental caries; and second, that excessive sucrose intake is bad for our general health or, at any rate, that of some particularly susceptible individuals.

Taking dental caries first, the presumptive evidence against sucrose is strong.[16] The British Dental Association dramatises the position in its pamphlet 'Stop the Rot' when it relates the 4 tons of teeth extracted annually from British children to the consumption of 585,000 tons of sweets. Prolonged exposure of developing teeth to sucrose-containing products is associated with rampant dental caries.[17]

Individually, changes in feeding habits are necessary to reduce the duration of exposure of the teeth to sugary foods—

this is the basis of the 'don't-eat-sweets-between-meals' advice. The manufacturer can sometimes modify the texture of sugary foods to make them less viscous and hence less likely to adhere to the teeth and in this way reduce exposure time. Finally, the food might be made free from sucrose. This more immediately poses the problem of what to use instead.

Where bulk and texture is relatively unimportant the non-caloric sweeteners cyclamate and saccharin can be used. Their use in baby and infant foods, however, contravenes the advice of the FAO/WHO Joint Expert Committee[18] and so the manufacturer must replace sucrose with other less cariogenic carbohydrates. To do so creates shelf-life problems, for the least cariogenic carbohydrate, starch, has the least osmotic preservative value. Successful adoption of a sucrose-free formula unaided by artificial preservative, demands, as we have found in developing a sucrose-free vitamin C syrup, changes in manufacturing technique and even higher standards of microbial cleanliness than those already customary.

Before developing this theme in relation to foods, let me say that it seems to be relatively easy for independent research workers to tell the manufacturer what he ought to do but rarely can they tell him how to do it. Moreover, they are slow to give him effective support when he takes a few steps in the right direction, yet such is the nature of good innovation. As Bacon said in his 'Essay on Innovation', published around 1612, 'It were good if men in their innovations would follow the examples of time itself, which indeed innovateth greatly but quietly and by degrees scarce to be perceived'.

If it should become generally accepted medical opinion that on public health grounds the consumption of sucrose should be restricted, the introduction of sucrose-free foods will necessarily be slow because new techniques of formulation, processing and perhaps preservation will have to be developed.

Relative to sucrose, sweetness is provided more cheaply by non-nutritive sweeteners but because they contribute nothing but sweetness they cannot substitute directly for sucrose in most products. Alternative carbohydrate sweeteners are considerably more expensive than sucrose as a source of sweetness. In some types of food a combination of non-nutritive sweeteners with one of the many starch hydrolysis products, including the usual end-product glucose, may prove the best alternative to sucrose available to us. By varying the hydrolysis conditions for starch, and by using specific enzymes to effect conversion, a variety of nutritive glucose syrup (liquid glucose) products can be made which vary widely in sweetness, viscosity and flavour. All of them are sucrose and fructose-free and many have a high proportion of relatively large yet quickly assimilable and easily digestible molecules.[19] A recent, slight trend towards lower consumption of sucrose and higher consumption of glucose is discernible in the U.K.[15, 20]

SUCROSE AND DISEASE

Ischaemic heart disease and diabetes, two conditions in which excessive sucrose consumption has been implicated, are clearly multifactorial ailments and more and stronger evidence than that currently available will be needed before food manufacturers would be justified in abandoning the general use of sucrose—a step with important economic and political implications.

Long-term dietary studies in man are needed but such studies are difficult. Ideally the subjects should be locked up and I believe there is a strong case for permitting volunteer prisoners to participate in dietary studies.

It is a pity, though perhaps inevitable, that much of the experimental work to test the metabolic differences between sucrose and other carbohydrates has been done with relatively high, often unphysiological, amounts of sucrose, for relatively short periods of a few days or weeks. This work has, nevertheless, drawn attention to differences in metabolic effects, including the possibility that a sucrose calorie is more likely to be converted to fat than is a starch, glucose or liquid glucose calorie. This latter possibility is clearly of tremendous importance to food manufacturers and so we are studying it.

The first evidence is over 50 years old, for it was Higgins in 1916 who concluded, from study of changes in the respiratory quotient in man, that fructose, and hence sucrose, 'shows a tendency to change into fat in the body', as compared with dextrose.[21] We found evidence of this in rats when, after six months feeding of 80% carbohydrate diets to adult male rats, we observed that carcass and liver fat were significantly increased by sucrose and fructose but not by liquid glucose or dextrose, as compared with a standard laboratory diet[22] *(Table 7)*. In a subsequent pilot experiment on baboons we found more body fat in animals fed a 75% sucrose or fructose diet than in those fed liquid glucose, dextrose or starch.[23]

In another study recently completed with mature male baboons we again found an 80% sucrose or fructose diet to be associated with the deposition of significantly more body fat after six months than a liquid glucose diet[24] *(Table 8)*.

Table 7. EFFECT OF DIETARY SUGARS IN THE RAT

20 *per group*

26 *week diet* 80% *as*	*Carcass Fat* % *body weight*
Fructose	21·8*
Sucrose	23·2†
Dextrose BPC	19·7
Liquid Glucose BPC	19·8
Control Diet 41B (60% starch)	18·0

*significant at $0{\cdot}01 < P < 0{\cdot}05$
†significant at $P < 0{\cdot}001$

Table 8. EFFECT OF DIETARY SUGARS IN THE BABOON

Means and S.E. of Means—five per group

26 *week diet*, 76% *calories as*	*Energy value of diet consumed* Kcalories × 10³	*Omental, pericardial, retroperitoneal fat Wet weight as* % *final body weight*
Fructose	151·21 ± 5·97	1·153 ± 0·350
Sucrose	149·44 + 5·31	1·655 ± 0·277
Liquid Glucose BPC	154·14 ± 5·11	0·514 ± 0·131

Given the genetic variability of baboons caught in the wild, their unknown ages and previous dietary history, it is surprising that in only six months significant differences in body composition appear which are apparently related to the type of carbohydrate fed.

With man, we have some evidence from work in progress that a fall in fasting serum cholesterol concentration might be brought about in middle-aged men who habitually use table sucrose, merely by replacing their table sucrose isocalorically with glucose. If this effect can be consistently demonstrated in repeat experiments, there would be grounds for believing that in man, as in baboons and rats, glucose is a less fattening carbohydrate than sucrose.

The food manufacturer is the agent through which research evidence relevant to human nutrition and dental health is translated into changes in product composition. It is now clear that different dietary carbohydrates may have different metabolic effects in the human body and may affect the course of dental disease in different ways. Nevertheless, much of the evidence has been obtained from short-term experiments with diets of abnormal composition. Alert food manufacturers are aware of the research going on and will take account of it first in the formulation of products for special dietary uses. Subsequently, using the experience gained, more general changes in the types of carbohydrate used in food manufacture could follow. However, as I hope my paper has shown, sucrose replacement in food on a large scale is technologically a formidable operation and with it are associated other considerations, economic and political in nature.

REFERENCES

1. KARE, M. R. and MALLER, O., *The Chemical Senses and Nutrition,* John Hopkins, Baltimore, 198 (1967)
2. BROOK, M., *R. Soc. Hlth J.,* **89,** 140 (1969)
3. COTTON, R. H., REBERS, P. A., MAUDRU, J. E. and RORABAUGH, G., 'Uses of Sugar and Other Carbohydrates in the Food Industry', *Adv. Chem. Ser.,* **12** (1955)
4. *Cocoa, Chocolate and Confectionery Alliance Annual Report,* The Cocoa, Chocolate and Confectionery Alliance, 11 Green Street, London, W.1. (1967–68)
5. McHUGH, W. D., McEWEN, J. D. and HITCHIN, A. D., *Br. dent. J.,* **117,** 246 (1964)
6. *The Condensed Milk Regulations,* HMSO, London (1959)
7. *The Food Standards (Preserves) Order,* HMSO, London (1953)
8. NICHOLS, A. A., *J. Dairy Res.,* **10,** 231 (1939)
9. NISCOX, E. R., *Chem. Inds. Lond.,* **26,** 365 (1947)
10. ERICKSON, F. J. and FABIAN, F. W., *Fd Res.,* **7,** 68 (1932)
11. *The Soft Drinks Regulations,* HMSO, London (1964)
12. *The Food Standards (Ice-Cream) Order,* HMSO, London (1953)
13. *The Food Standards (Ice-Cream) Regulations,* HMSO, London (1959)
14. BUNKER, H. J., 'Dealing with Microbiological Troublemakers' in *Biology and The Manufacturing Industries,* Ed. by M. Brook, Academic Press, London (1967)
15. *Board of Trade Journal,* **194,** 753 (1968)
16. WINTER, G. B., *Br. dent. J.,* **124,** 407 (1968)
17. Panel on Cariogenic Foods, *1st Report,* Committee on Medical Aspects of Food Policy, Dept. of Health and Social Security (1969)

18. *Tech. Rep. Ser. Wld Hlth Org.*, **228,** 6 (1962)
19. ALLEN, R. J. L. and BROOK, M., *Am. J. clin. Nutr.*, **20,** 163 (1967)
20. *Board of Trade Journal*, **196,** 310 (1969)
21. HIGGINS, H. L., *Am. J. Physiol.*, **41,** 258 (1916)
22. ALLEN, R. J. L. and LEAHY, J. S., *Br. J. Nutr.*, **20,** 339 (1966)
23. ALLEN, R. J. L., BROOK, M., LISTER, R. I., SIM, A. K. and WARWICK, M. N., *Nature, Lond.*, **211,** 1101 (1966)
24. BROOK, M. and NOEL, P., *Nature, Lond.*, **222,** 562 (1969)

II

THE CHEMISTRY OF SUCROSE

Chapter Five

THE CHEMICAL REACTIVITY OF SUCROSE

LESLIE HOUGH
Professor of Chemistry, Queen Elizabeth College, University of London

Sucrose is produced in vast quantities, on a scale that is larger than for any other pure organic chemical, it is low in price, consequently strenuous efforts have been made over the past thirty years to devise chemical expedients for the exploitation of sucrose on an industrial, and hence economic, scale. These efforts have been fraught with many difficulties but they are not insurmountable, as for example the successful development of sucrose-based detergent.[1] Nevertheless, the problems encountered do reveal, and emphasise, the need for a major effort to be expended on basic research into the chemistry of this fascinating molecule of nature.

Sucrose is a non-reducing disaccharide of unique structure (I), numbered as shown, containing eight alcoholic groups, three of which are primary hydroxyls (at carbons 6, 1′, and 6′), the remaining five being secondary hydroxyls. Since the disaccharide is hydrolysed with extreme ease by acids but is reasonably stable in the presence of strong bases or alkalis, the preparation of sucrose derivatives is restricted to basic media. Partially substituted derivatives of sucrose are very difficult to isolate in a pure form because of the formation of mixed products, resulting from the multiplicity of hydroxyl groups present, although the hydroxyl groups do vary slightly in their rates of reaction. In theory, esterification with one molecular proportion of reagent could give rise to eight possible mono-esters of sucrose, whilst in practice the primary hydroxyls react preferentially, in particular those at carbons 6 and 6′, but the products are always contaminated with small amounts of other mono-esters and di- and tri-substituted esters, the proportions varying according to the reaction conditions and the reactants. The greater reactivity of the primary hydroxyls of sucrose is attributed to their less hindered (or more exposed) positions on the chair conformation (C_1^4) of the D-glucopyranosyl ring, and the envelope (E^4) conformation of the D-fructofuranosyl ring (II).[2]

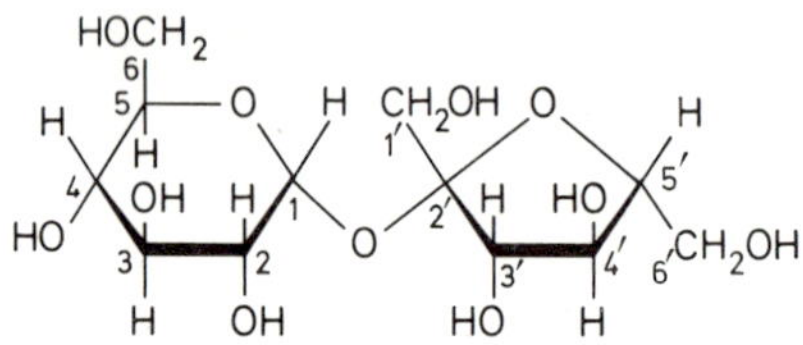

I

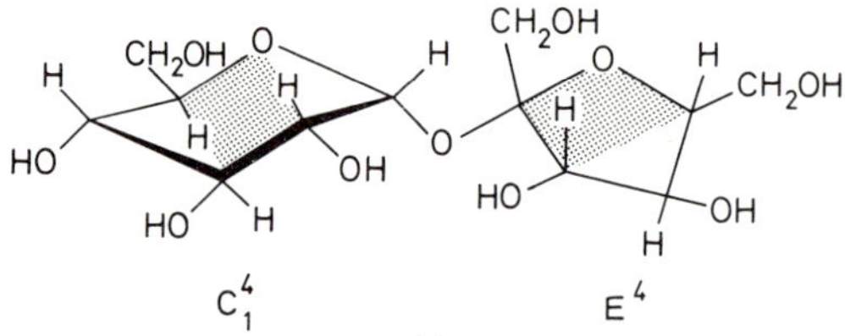

II

A variety of mono-esters of sucrose (III) have been prepared by trans-esterification under carefully controlled conditions, including the stearate, palmitate, myristate and laurate, and they are non-toxic, tasteless and as detergents have the advantage that they are neutral and bio-degradable.[1] They are made by reacting sucrose with methyl stearate, in the presence of potassium carbonate, with dimethylformamide (DMF) as solvent. The commercial availability of this solvent was undoubtedly a major advance in sucrochemistry because of the lack of suitable solvents hitherto.

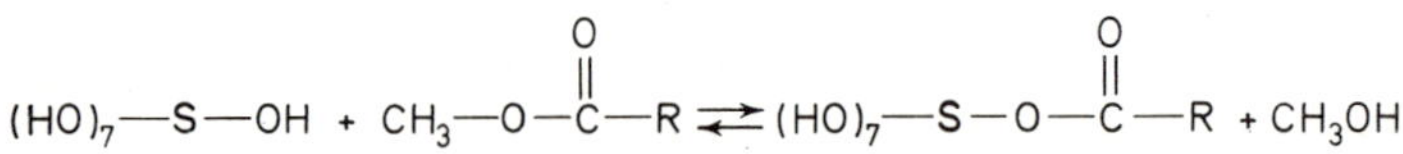

III

Initial studies on the chemical structure of the mono-esters indicated that the fatty acid was located on carbon-6 of the glucose unit but later studies by Lemieux's group[3] suggested that the 6′-position of fructose was preferred (62%), with substantial substitution at carbon-6 (28%), the remainder (10%) being at other positions. Subsequent studies by Reinefeld and Klaudianos[12] have shown that the 6-ester predominates, followed by the 6′-ester; amongst the diesters, the 6, 6′-diester predominates. Structural work on the esters is complicated by the possibility of migration of ester groups, via cyclic ortho-esters, from one hydroxyl to another under the influence of bases. In a few cases only have detailed structures of partially substituted sucroses been determined. In 1935, Percival[4] showed the higher reactivity of the primary positions of sucrose in etherification, when he obtained 6, 1′, 6′-tri-*O*-methylsucrose from the methylation of potassium hydroxide–sucrose addition compound ($C_{12}H_{22}O_{11}$, 3KOH) with dimethyl sulphate. In a related study, Black, Dewar and Rutherford[5] reacted methyl toluene-*p*-sulphonate with a tri-sodium derivative of sucrose, prepared by reaction of sodium or sodium amide with sucrose in liquid ammonia at −33°, and obtained a complex mixture of methyl ethers (OCH_3 content, 10·6%). Fractionation by chromatography gave a mixture of six mono-*O*-methylsucroses; careful analysis suggested that substitution predominated at position 2 and to a lesser extent at position 6 of glucose, and at position 1′ of the fructose closely followed by position 3′. These results closely parallel those obtained on methylation of the glucopyranosyl units of alkali–cellulose with methyl iodide, when the rate constants at positions 2, 3 and 6 were in the ratio 5:1:2 respectively.[6] Close proximity to the anomeric centres (carbon 1 of glucose and carbon 2′ of fructose) is clearly a major factor in these etherification reactions.

An alternative approach to the synthesis of partially substituted sucrose derivatives was developed during a study of the preparation of alkoxycarbonyl derivatives.[7] Thus gradual addition of aqueous alkali to a cooled, stirred mixture of sucrose and an alkyl chloroformate

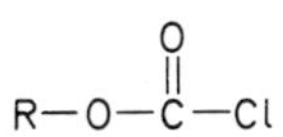

gave a water-insoluble syrup which separated in high yield from the reaction medium. The product is a complex mixture of *O*-alkoxycarbonylsucroses containing, on average, 5–6 alkoxycarbonyl groups per molecule of sucrose.

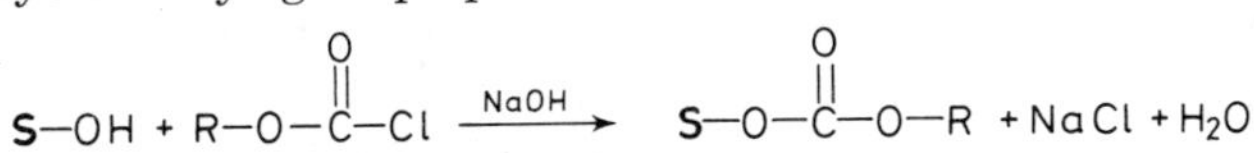

When light petroleum was added to the reaction mixture, a tri-alkoxycarbonylsucrose derivative was precipitated from the medium: under these conditions and using ethyl chloroformate, a 75% to 80% yield of syrupy tri-ethoxycarbonylsucrose was obtained. On heating this product at 150° under reducing pressure, it polymerised in 15 min to give an insoluble, resinous poly(sucrose) carbonate. The thermosetting polymer is formed by trans-esterification between ethoxycarbonyl groups and free hydroxyl groups with the elimination of volatile diethyl carbonate and ethanol *(Figure 1)*. The sucrose units are linked together by carbonate diester bridges to give a giant molecule. A bifunctional chloroformate ester, ethylene dichloroformate, or carbonyl chloride (phosgene) react with sucrose in the presence of a base to give amorphous, white, insoluble polymers of the polycarbonate type directly.

The extreme acid sensitivity of the inter-glycosidic linkage in sucrose is considerably strengthened by a small number of *O*-alkoxycarbonyl substituents, presumably by steric

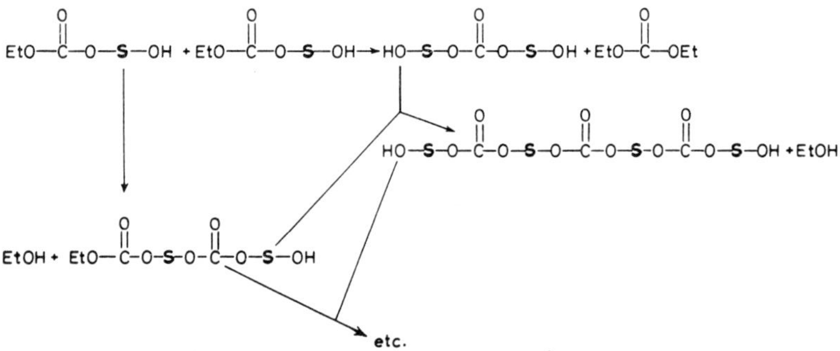

Figure 1. The polymerisation of ethoxycarbonylsucrose

inhibition in the transition step. A tri-*O*-ethoxycarbonylsucrose possesses higher acid stability than sucrose octaacetate *(Figure 2)*, whilst the octa-*O*-ethoxycarbonylsucrose is stable to concentrated H_2SO_4.[7]

In recent years we have concentrated our attention on the chemical synthesis of new derivatives of sucrose using sulphonate esters of sucrose (R-SO_2-OR′, where R = sucrose) as intermediates. The objectives of this study were three-fold, *(a)* to study the chemistry of new sucrose derivatives and so recognise the chemical properties and possible application, *(b)* to prepare bifunctional derivatives of sucrose with a view to the subsequent synthesis of linear polymers and *(c)* to study the selectivity and different chemical reactivity at the various positions of the sucrose molecule.

Lemieux and Barrette[8] found that when sucrose is treated with three molecular proportions of toluene-*p*-sulphonyl chloride in pyridine, it yields a mixture of penta-, tetra-,

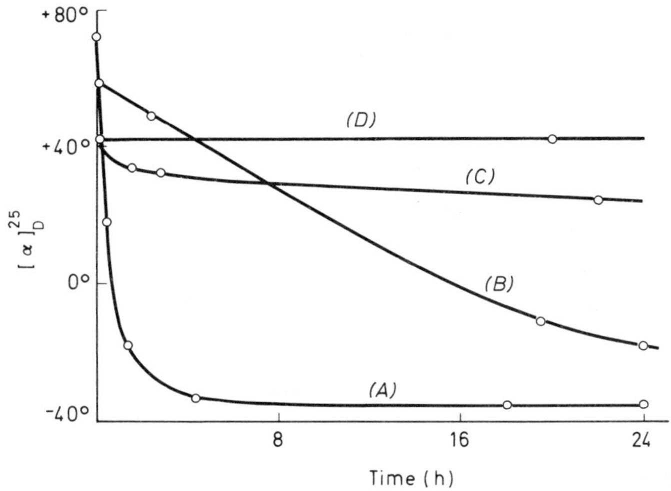

Figure 2. Hydrolysis of sucrose and its esters in N*–aqueous-alcoholic hydrochloric acid. (A)* Sucrose. *(B)* Sucrose octa-acetate. *(C) O*-Ethoxycarbonylsucrose (D.S., 3.1) *(D) O*-Ethoxycarbonylsucrose (D.S. 4.9)

tri-, and di-substituted products in the proportions 0·05, 0·3, 1·0 and 1·0 respectively. After chromatography the 'tri' fraction was shown to contain mainly syrupy 1′, 6, 6′-tri-*O*-tosyl-sucrose (VIII) and the 'di' fraction crystallised in about a 20% overall yield to give 6, 6′-di-*O*-tosyl-sucrose (IV). Treatment of this ditosylate (IV) with sodium methoxide in methanol gives the crystalline, 3, 6-3′, 6′-di-anhydro-sucrose (V).[9] In this reaction the sucrose molecule undergoes a conformational change (VI→VII). The shape of the glucose and fructose rings alter and this modifies the chemical properties of the molecule; the hydroxyl groups of the glucopyranosyl unit now occupy more crowded axial positions in (VII) in contrast to more exposed equational positions in sucrose—as in (VI).

Lemieux and Barrette[8] treated the 'tri' tosylate of sucrose (VIII) similarly and obtained a 1′, 2; 3, 6; 3′, 6′-tri-anhydro-sucrose (IX), a bridged sucrose derivative with only two of

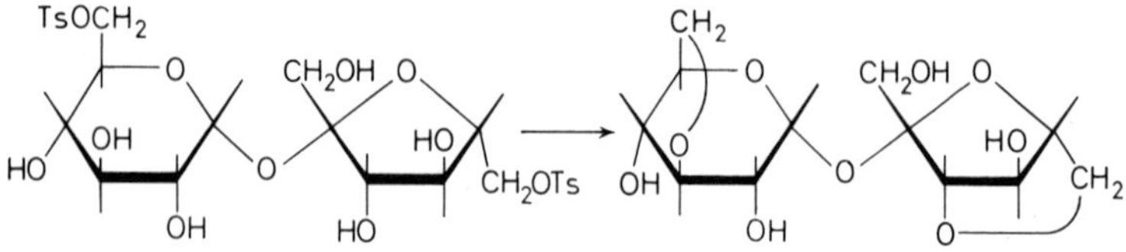

IV (Ts = CH_3—C_6H_4—SO_2—)

V

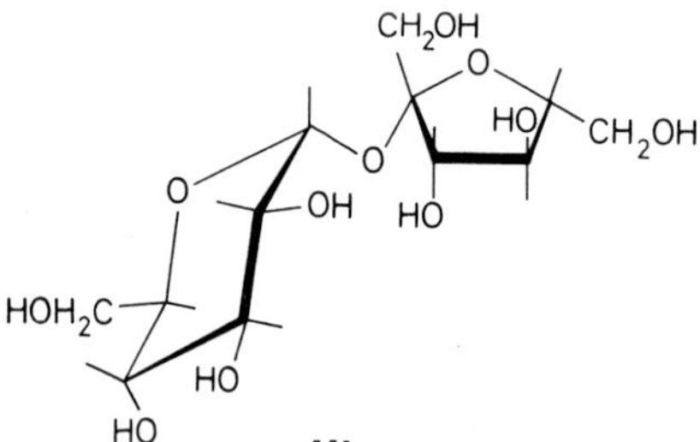

VI

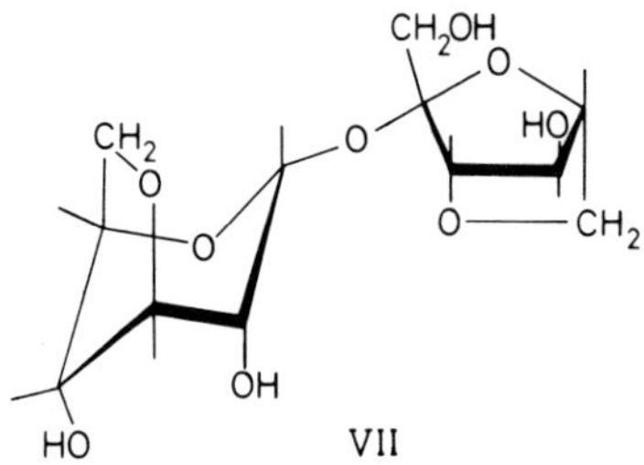

VII

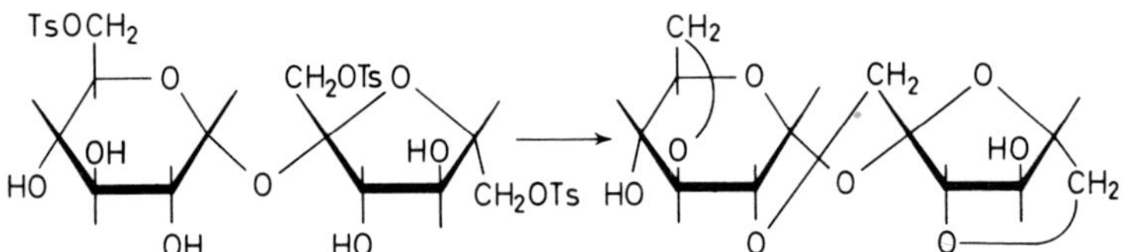

VIII

IX

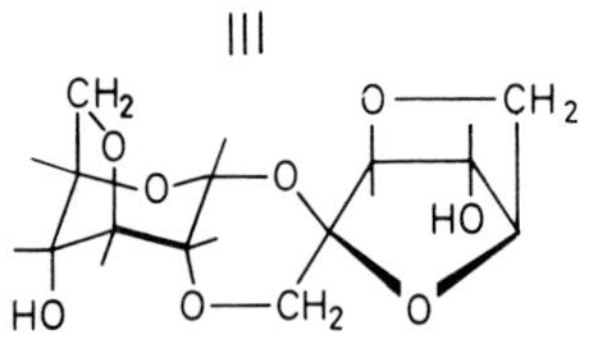

III

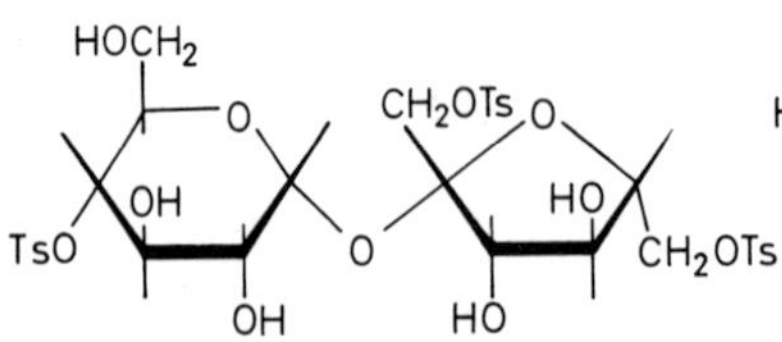

X

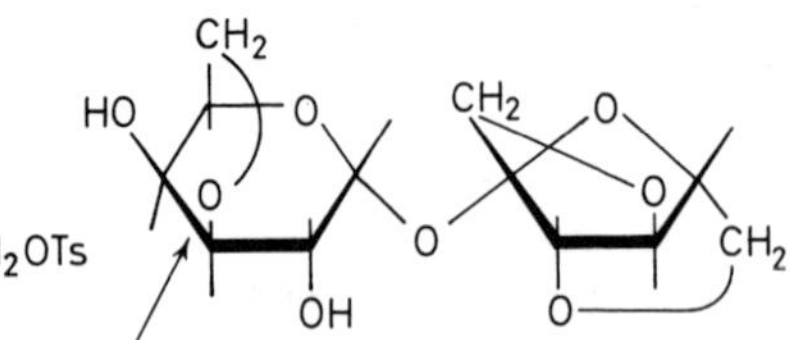

Galacto

XI

the original eight hydroxyls remaining exposed. The Canadian workers[8] also prepared the 4, 1′, 6′-tri-*O*-tosylate of sucrose (X) which was transformed with sodium methoxide into 3, 6-anhydro-α-D-galactopyranosyl-1′, 4′-3′, 6′-di-anhydro-β-D-fructofuranoside (XI), via a 3, 4-epoxide and then a 2, 3-epoxide, a remarkable series of transformations in which the configuration at carbon 4 of sucrose has been inverted (gluco → galacto).

We have prepared a variety of 6, 6′-bifunctional derivatives (XII) of sucrose in high yield, by exploiting the reactions of the crystalline 6, 6′-di-*O*-tosyl-sucrose (IV) with various nucleophiles (x^-), the incoming nucleophile replacing the sulphonyl-oxy groups.[9] These reactions can be duplicated using the octa-*O*-mesylsucrose, when the primary positions at 6 and 6′ react preferentially. This research was facilitated by (*a*) thin layer chromatography (T.L.C.) for following the

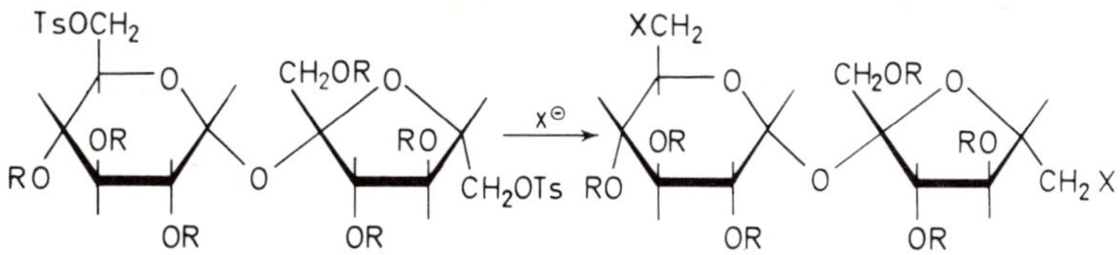

IV XII (X = Cl, Br, I, N_3, SCN)

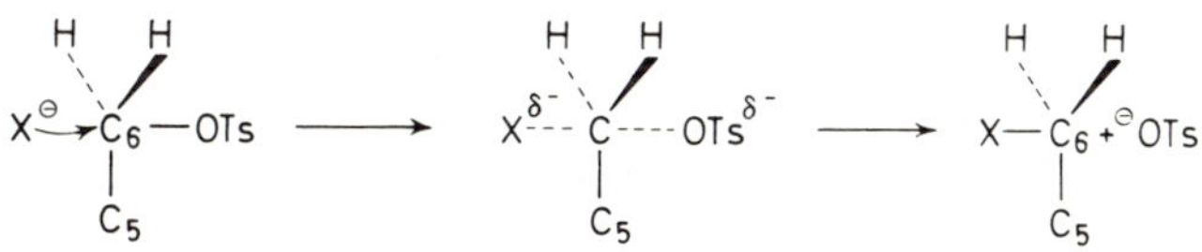

Bimolecular Nucleophilic Substitution

reactions in order to determine the optimum conditions and by-products, (*b*) aprotic solvents of high dielectric constant, which favour the nucleophilic substitution reactions—these solvents include, dimethyl formamide (DMF), dimethyl sulphoxide (DMSO) and hexamethylphosphoric triamide (HMPT)—and (*c*) nuclear magnetic resonance (NMR) for structure determination. Using one of these bifunctional derivatives, we prepared a 5, 5′-diene derivative of sucrose (XIV) by treating 6, 6′-dideoxy-6, 6′-di-iodo-sucrose (XIII) with silver fluoride in pyridine. In this diene (XIV) the asymmetry at carbon 5 and 5′ is destroyed but hydrogenation reconstitutes the asymmetry giving four 6, 6′-dideoxy derivatives, including the sucrose derivative (XV), and the L-ido, L-sorbo analogues.[10] The diene hexa-mesylate (XVIII) is readily accessible from the octamesylate (XVI) which reacts selectively with sodium iodide in ethyl methyl ketone to give the 6, 6′-di-iodo derivative (XVII), crystalline and in high yield. Proof of structure of the latter followed from mesylation of the 6, 6′-ditosylate (IV) to give the hexamesylate (XIX) and reaction of the latter with iodide to effect selective substitution of the tosyloxy substituents giving the 6, 6′-di-iodo derivative (XVII). The NMR spectrum of 6, 6′-dideoxysucrose (XV) produced by hydrogenation of the di-iodo derivative (XVII), showed two doublets, resulting from the

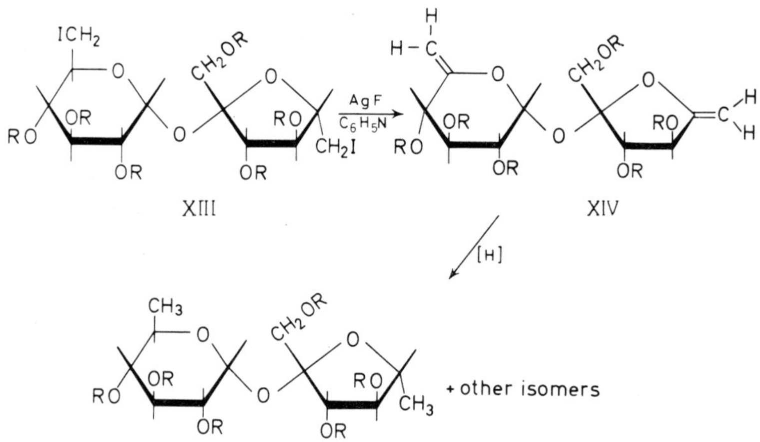

XV

splitting of the methyl signals by the adjacent hydrogens at carbon 5 and 5′ respectively, thus confirming the structure. These experiments show quite conclusively that, in nucleophilic substitution reactions, the 6, and 6′ positions of sucrose are far more reactive than the other primary 1′ position, a major factor in sucrose chemistry that has not been appreciated despite substantiating evidence in fructofuranose derivatives.

During the benzoylation of sucrose 6, 6′-ditosylate (IV) it was noted that the reaction must be carefully controlled, at 0°, to avoid the formation of by-products. At room temperature, 6, 6′-dichloro and monochloro derivatives are produced: these products can be produced in good yield by reaction of the ditosylate (IV) or the octamesylate (XVI) with pyridine hydrochloride or sodium chloride in DMF. Similarly crystalline mono-, di- and tri-iodo derivatives have been isolated from the reaction of the octa-mesylate (XVI) with sodium iodide in HMPT. Using sodium bromide in HMPT, crystalline mono-, di- and tri-bromides have been isolated.

The isolation of mono-halogeno derivatives of sucrose shows that either the 6 or 6′ position is more reactive than the other

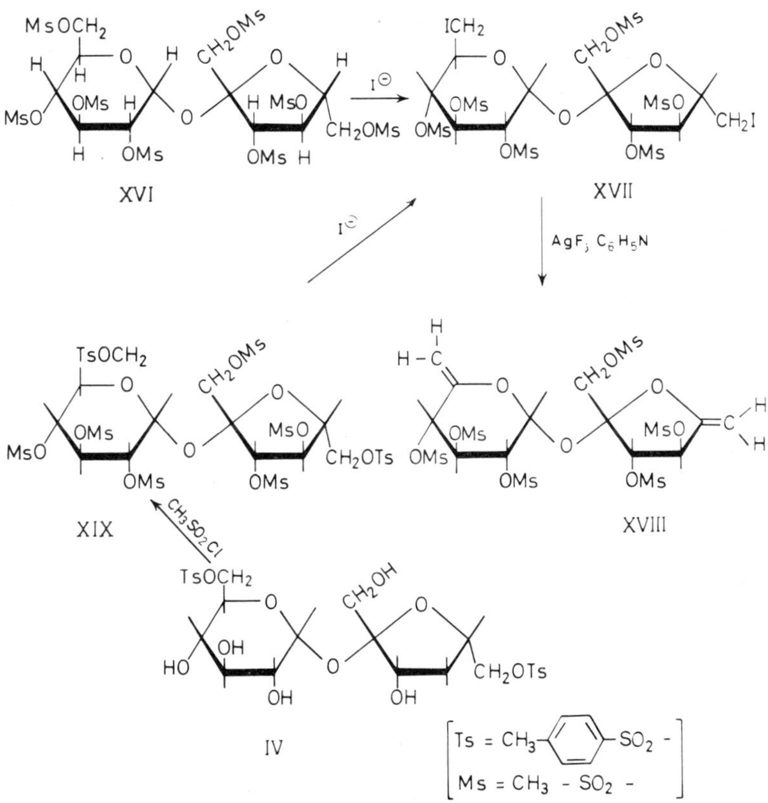

in these nucleophilic substitution reactions. Characterisation of the mono-substituted products is in hand.

Another approach to the direct synthesis of chloro derivatives from sucrose was described by J. K. N. Jones' group who found that sucrose reacted with sulphuryl chloride in pyridine to give a crystalline 1′, 4; 6, 6′-tetrachloro-tetradeoxy 'sucrose' di-cyclic sulphate (XX).[11]

In summary, methods are now available for the synthesis of mono-, di-, tri- and tetra-functional derivatives of sucrose, including deoxy-, halogeno-, thio- and amino-derivatives. A fundamental knowledge of the chemistry of the sucrose

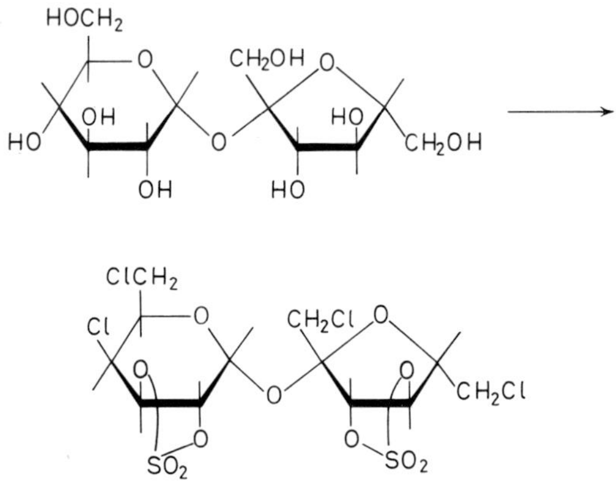

XX

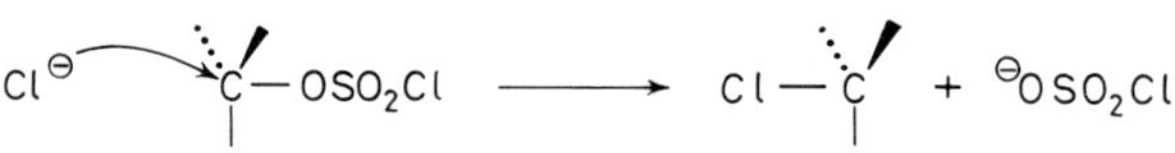

molecule, widely accepted as one of the most difficult and challenging topics in the carbohydrate domain, is gradually emerging.

We are grateful to the Sugar Research Foundation for their continued support of our sucrochemical studies initially at Bristol University and later at Queen Elizabeth College, and to my colleagues, Dr. R. S. Theobald, Dr. C. Bolton, Dr. R. Khan and Dr. K. S. Mufti, who carried out various aspects of the sucrochemistry described above.

REFERENCES

1. 'Sucrose Ester Surfactants', *Research Report,* Sugar Research Foundation Inc., New York (1961) Literature citations (1963)
2. RODD, E. H., *Chemistry of the Carbon Compounds,* Elsevier, **IF,** 114, 125 (1968)

3. LEMIEUX, R. U. and McINNES, A. G., *Can. J. Chem.*, **40,** 2394 (1962)
4. PERCIVAL, E. G. V., *J. chem. Soc.*, 648 (1935)
5. BLACK, W. A. P., DEWAR, E. T. and RUTHERFORD, D., *J. chem. Soc.*, 3073 (1959)
6. SUGIHARA, J. M., *Adv. Carbohyd. Chem.*, **8,** 1 (1953)
7. THEOBALD, R. S., *J. chem. Soc.*, 5359, 5365, 5370 (1961)
8. LEMIEUX, R. U. and BARRETTE, J. P., *J. Am. chem. Soc.*, **80,** 2243 (1959) *Can. J. Chem.*, **37,** 1964 (1959); **38,** 656 (1960)
9. BOLTON, C., HOUGH, L. and KHAN, R., *unpublished results*
10. HOUGH, L., KHAN, R. and OTTER, B. A., *Adv. Chem. Ser.*, **74,** 120 (1968)
11. BRAGG, P. D., JONES, J. K. N. and TURNER, J. C., *Can. J. Chem.*, **37,** 1412 (1959)
12. REINEFELD, E. and KLAUDIANOS, S., *Zucker*, **21,** 330 (1968)

Chapter Six

UTILISATION OF SUCROSE BY THE CHEMIST

JOHN L. HICKSON

Vice President and Director of Research, The International Sugar Research Foundation, Inc., Bethesda, Maryland, U.S.A.

Refined sugar has been a significant factor in international trade for more than six centuries which, as you can see, more than spans the entire evolution of the organic chemicals industry. One can only imagine how many attempts have been made to harness this abundant, low-priced, pure disaccharide as a raw material for chemical applications. Even so, statistics of the United States Department of Agriculture disclose that less than 2% of the 10·5 million tons of sugar delivered in 1968 went into non-food uses. But, as you may not realise, the biggest non-food outlets are in the pharmaceutical and tobacco industries.

You may well wonder why such an obvious organic chemical raw material available for so long has achieved so little chemical utilisation. Could it be due more to an intransigence of the molecule or to faint-heartedness of the chemist? I may well confide in you. I believe it to be more the fault of the molecule than the chemist.

Professor Hough documents the sophistication of the chemistry required to successfully modify sucrose at known and predicted locations *(see Chapter 5).* Even so, since the middle of the nineteenth century when a rational organic chemistry started to evolve, almost all of the standardised reactions, oxidation, hydrogenation, treatment with strong acid or strong base, dehydration, etc., have been demonstrated to yield useful materials from sucrose by degradation. Almost invariably, however, it has been discovered that other raw materials lower in cost turn out to be more economical sources of these commodities and sugar has not gained the market. Examples of such products include hydroxymethylfurfural and levulinic, oxalic, mucic and lactic acids. Only recently have such sophisticated processes as the hydrogenolytic production of glycerol from the hexitols derived from invert sugar, been demonstrated to be successful competitors for petrochemical processes. Even so, the products of such technologies create markets for only millions of pounds of sucrose while food outlets are measured in terms of millions of tons. What chance does sucrochemistry offer in its quest for relevance? Certainly, in view of a century of negative responses, it could seem folly that the sugar industry continues to pursue this will-o'-the-wisp.

Perhaps through my subsequent documentation, you will come to some understanding of my optimism for the future of sucrochemistry.

Sugar is almost ubiquitous. It is produced by all known land plants and is recovered commercially from a root (44%), several grasses (56%), and a tree (<1%). Some countries are principally exporters, some mainly importers, and most rely partially on domestic sources and import the remainder of their requirements. Of the 77 million short tons produced in 1968, about 30% moved in international commerce. It's a big business.

Twenty-six years ago, the U.S. producers recognised that food outlets of the westernised populations had become static. Indeed, the *per capita* disappearance has not increased for nearly 50 years (contrary to some fallacious allegations). The dramatic growth of the western markets occurred up to the time of the first world war. In fact, the current increase in sugar disappearance in the western world is due to more consumers eating the same amount. It is the emergence of new, vigorous national economies of the less developed nations that have been swelling the *per capita* disappearance figures on a world basis.

The U.S. sugar suppliers banded together in 1943 to give industry-wide support to utilisation research on the commodity, sucrose. They were joined by British Canadian, Irish, European, Australian, and South African producers. Over the 26 years, they have expended more than $5·9 million through 290 grants to universities and research institutions to solve their problems and improve their destiny. Of these 290 grants, 97 supported studies on non-food uses. Of these, 78 were directly in sucrochemistry. This information, in its entirety, has been reviewed in the manuscript of a book, 'Sucrose

Chemicals', which should appear sometime in 1970. I propose that we step through the looking glass into the arena of the marketing of organic chemicals. Here in the light of the chemical properties of sugar, I will share with you our current thinking with respect to where markets might be found and something about the mechanisms employed by the Foundation to try to achieve them.

There are some sixteen markets for organic chemicals of such a size that successful sucrochemistry could dispense significant amounts of sugar. These are:

(1) Foods
(2) Feeds
(3) Fuels
(4) Explosives
(5) Elastomers
(6) Lubricants
(7) Solvents
(8) Soil conditioners
(9) Fibres
(10) Adhesives
(11) Paper
(12) Pesticides
(13) Plasticisers
(14) Plastics
(15) Surface coatings
(16) Surfactants

(1) Since *Food* constitutes the great preponderance of the current market, we will consider it to be a special subject outside the boundaries of this discussion.

(2) *Animal Feed* markets in the Atlantic community are available only in special situations, for the production of the grain and forage fields in temperate climes produce feed energy at much lower prices. At times, distress priced sugar is available, but not always. Also, there do appear some special purposes such as enhancing the acceptability of special feeds, as in 'starting rations' for baby pigs or in experimental studies where, for special purposes such as trace food-factor analysis, one must exclude from the diet starch which normally carries with it significant amounts of B vitamins, phosphates, etc. However, until the demand for grains for food raises their price by factors of, perhaps, two to three, we cannot expect a large outlet for sugar in these feeds.

(3) Some 51% of the weight of sucrose is oxygen. This essentially removes sugar from consideration as a *Fuel*.

(4) Small amounts of sugar are made into *Explosives* as the octanitrate in nitroglycerine lines and some, as the second derivative, mannitol hexanitrate, but these are low poundage outlets except in times of major conflict.

(5), (6) The eight hydroxyl groups result in such high levels of hydrogen bonding that sugar probably has little market potential either in *Elastomers* or *Lubricants*.

(7) One can blame sugar's cost and its molecular size for

giving us little hope in *Solvents* in the western world. But, remember that alcohol from fermented molasses is one of the major sources in the less developed countries.

(8) We have not yet found ways of combining sufficient numbers of sucrose molecules into the highly polymerised structures to overcome sugar's inherent biodegradability, so we predict not much future in *Soil conditioners*.

(9) Sugar's eight hydroxyl groups, of which four or five have very similar reactivity, give little substance for hopes of obtaining the long linear polymers that are requirements for *Fibres*.

Thus, you see we have disposed of nine of the sixteen items in the above list; but there are seven areas in which we have made some extended research forays.

(10) In *Adhesives*, we have not worked with the protein glues, but there is a patent literature. We have modified ureaformaldehydes—to find no advantage. We can extend the melamine formaldehydes for special purpose adhesives with real advantages. In the very big plywood adhesives field, we can make some significant contribution to phenol formaldehyde. The total amount of the Foundation's work in these areas is, so far, rather limited.

(11) In *Paper*, we have not extended our study of the melamine formaldehyde resins to the paper field. We have studied textile applications and others, but we do not know what these products would do for improving the wet strength of towelling and packaging materials.

(12) Concerning *Pesticides*, some years ago we pursued for a few years a hypothesis that if one were to attach the toxic moieties of a number of known pesticide molecules on to sucrose through ether or ester linkages, one might gain water dispersibility and, perhaps, even cell permeability. The results were not useful.

(13) Concerning *Plasticisers*, our programme has produced a number of derivatives, such as combinations of the long chain fatty acid esters with those of small organic acids. Two of these have been commercialised—one, sucrose diacetate hexaisobutyrate; the other, sucrose octabenzoate. They are being marketed for plasticising ester-type polymers, e.g., the cellulosics. The really big current plastics, however, the polyethylenes and the vinyls, have eluded our grasp. This is one of the research goals we hope to reach.

(14) In *Plastics*, the Yarsley Laboratories review a signi-

ficant amount of the work that we have carried on in this area (*see* page 69). We think much of the opportunity now is in merchandising the know-how already in hand.

We have a concept that there are much greater opportunities than have been tapped thus far. We expect to apply the basic research that has been evolving at Queen Elizabeth College under Professor Hough. He has created a sequence of a score of derivatives of sugar with reactive groups at known locations, principally carbons 6, 6′ or 1′. Our plan is to use these new reactive centres to combine sugar, through copolymerisation or graft polymerisation reactions with a variety of other fibre forming, film forming, or bulk plasticising polymers to modify the hydrophilicity of the surface. I believe we are almost at the point where this is a necessary resultant of the evolution of our understanding of polymerising processes.

My enthusiasm arises from a glimpse of the kinds of things that might be done. Hydrophilic polymers are used in human prostheses for eyes, for replacing cartilage in plastic surgery, or, at least possibly, in surgical palliation of arthritis. They could be most desirable for the membranes in reverse dialysis equipment used in recovering life-bestowing water from brackish sources. They have great commercial potentials in modifying nylons and terylenes so widely used in textiles.

In America, much of the automobile upholstery is spun, extruded, or calendered from vinyl polymers. In our dry winter climate, to slide across the seat getting under the steering wheel can create enough static electricity to draw a half inch spark when you put the key in the ignition. Each winter, I am electrifyingly reminded how grand an achievement it would be to create a sucrose-containing surface coating for the plastic on that front seat so that I would avoid a knuckle-busting shock on that first cold, crisp morning.

Dental prostheses for the most part are moulded from acrylate resins. Hence, vinyl or allyl ethers or acrylate esters of sucrose might alter the properties enough to make them less uncomfortable in the mouths of unfortunates who require this kind of help.

Suffice it to say that in the minds of us at the International Sugar Research Foundation, the fields of plastics and plasticisers loom as two of our major targets for sucrochemical endeavours.

Please note how much importance we attach to the work by Professor Hough and his students and how it dovetails into this overall concept. Also, please note how significant it is to have the skills and ingenuity of applications groups as exemplified by the Yarsley Laboratories.

(15) Two decades ago, a study at the Eastern Regional Laboratories disclosed that allyl chloride could be reacted with sucrose in the presence of an aqueous alkali to produce a triallyl derivative. This product, in several lacquer establishments, was discovered to be a really first rate *Surface coating* for wood. At about this time, my predecessor, Dr. Henry B. Hass, took over the Foundation's research programme. Dr. Hass, always acutely aware of the economic imperatives, re-examined the research results to discover the process involved an almost 100% conversion of the 10¢ sugar, but utilised only half of the 35¢ allyl chloride. Needless to say, this thrust the cost of this desirable product so high that markets never evolved for it. However, it did demonstrate that suitably modified sucrose displayed some very desirable properties in surface coatings.

Some years after that, after having taken our foray into surfactants, as I shall tell you a bit later, we applied the transesterification reaction technique to the flax, tung, and soya bean oils that are the common vehicles of the paint industry. We discovered, as Dr. Hass had predicted, that the increased functionality of the basic molecule conveyed with it the expected enhancement of the polymerisability. To illustrate my thesis, you may consider that methyl-linoleate does not dry at all. The linoleate ester of ethylene glycol never dries beyond a very tacky stage. Linseed oil, the tri-linoleate of glycerol, has been the standard drying oil of the paint industry for many centuries. If the paint chemist wants to increase his functionality beyond this, he chooses such polyols as sorbitol, with which he gets a tetra-ester (it dehydrates to a sorbitan in the kettle) or pentaerythritol. Here he also gets a tetra-ester except that he has increased the cost of his polyol into the 25–30¢ a pound range.

Sucrose has eight hydroxyls, of which six or seven are readily accessible for reaction with linoleic acid by trans-esterification reactions. If he should use sucrose, the paint chemist could, as you can see, essentially double his functionality per molecule. He will not, nor should you, forget the cost of his polyol; sucrose then falls in the range of 10–12¢ a pound rather than the 25–30¢ if he employs equivalent derivatives from the petrochemical industry.

I have expounded on this theme to illustrate another of the

traits of the sucrose molecule that make it a likely candidate for extended commercialisation in sucrochemical outlets. Under plastics, I spoke of its hydrophilic character, an attribute desired in many polymer outlets. In surface coatings, I have emphasised its polyfunctionality at a low price.

With such observations, you are probably wondering why the paint industry is not buying tons of sugar. Well, we just have not finished our selling job. Here's why. About the time it looked as if the picture could be filled in, we were attacked by a competitor, the chemical business. It produced a molecule cyclohexylamine sulphamate, sold it at a tenth the price of sugar, seized upon controversial claims of dietary significance, and frightened the sugar industry into retaliation. Now that four years of research seem to have provided the basis for protection of the public from this hazardous additive, the Foundation prepares to gird itself for a return to more positive and productive research programmes.

The drying oil esters represent one of the areas of development to which I expect to return our emphasis.

One could examine the many other types of molecules applied in the surface coatings field. Phenol formaldehydes have been used widely as coatings in electrical insulation applications. Sugar can fit there. Phenolics are being replaced, to a large extent, by silicones, but we can make silyl derivatives of sucrose quite handily. We just have not begun to exhaust the possibilities in this area.

In the field of polymers, I neglected to cover the urethanes, but they fit in as well in surface coatings. The Foundation never did more than cast an envious eye at the urethanes because we concluded that the applications technology was so much more significant than chemical sophistication. As a result, we have expended our limited resources on other technologies and applauded while such companies as Pittsburgh Plate Glass, Union Carbide, ICI, and others have evolved urethane technologies involving the use of considerable amounts of sugar in the production of rigid urethane foams. Indeed, several of the major lines of refrigerators in the United States are truly 'smaller on the outside and bigger on the inside'. Economies in space truly are made possible by the greatly improved thermal insulating properties of this composite. The foam is said to be highly nonfrangible, another testimonial to the value of the increased functionality of sucrose where it can be utilised for its maximum impact.

Another big outlet for urethanes is the production of surface coatings as non-foamed formulations. We do not know much about it, but it is an area which probably we shall want to explore.

Perhaps you can conclude that, together with plastics and surfactants, surface coatings represent one of the fields with the highest potential for successful sucrochemistry.

(16) Finally, let us consider the topic *Surfactants*. Some of you are aware that in 1952, Dr. Hass cogitated about the idea that one of sucrose's major properties is the very strong hydrophilicity. He reasoned that if one could economically attach a fatty hydrocarbon tail to sucrose, he could produce the essence of a surface active molecule. Within two years, Foster D. Snell had produced a fatty acid ester which, indeed, was a surfactant. There followed perturbations and ramifications too numerous to relate. A score of companies licensed the processes and most of them lost their interest. But, not all did! A food grade surfactant is being produced in tonnage lots by the Dai Nippon Sugar Manufacturing Company in Japan. A surfactant applied in photographic emulsions is marketed in Germany by Farbenfabrieken Bayer. A French licensee, Melle-Bezons, forecasts significant marketing opportunities in reconstituted milks for calf feeding. Another sugar company, Raffinerie Tirlemontoise of Belgium, evolved a process apparently not in conflict with the Foundation patent claims which may have attractive marketing potentials in pulping applications. An American licensee for some years evolved its processes and products to the stage where they might be accepted by the United States Food and Drug Administration as surfactant additives for foods, only to withdraw from the development for the time being. Even so, from personal contact, I can assure you that the issue is very much alive in the corporate thinking of major producers of soap and detergents and companies having certain speciality markets.

In an actual homesite demonstration, detergents from sugar ester were demonstrated to be as biodegradable as soap. Both are a phase step away in biodegradability from the linear alkylates produced by the major petrochemical producers. I was told by a major officer of a significant U.S. detergent company that in less than ten years he expects enactment of quite restrictive legislation on water quality in the U.S. This will demand a new generation of detergent products. Sugar esters seem destined to be economically available at prices competitive with other candidates for this market. Yes, surfactants represent a major sucrochemical target for the Foundation.

SUMMARY

The organic chemicals business seems to offer several very real marketing opportunities for sugar limited only by the evolution of appropriate chemical modifications to fill definable objectives. The evolution of the basic chemistry of the molecule through current sophisticated techniques seems to be suggesting that these properties can be achieved.

Of the sixteen major areas of marketing of organic compounds, the fields of plastics, plasticisers, surface coatings, and surfactants seem to be most amenable to our developmental and, ultimately, marketing efforts.

Dr. Robert L. Bateman, Technical Director, Chemicals, of the Chase Manhattan Bank in New York, recently reviewed for the Chemical Marketing Research Society his outlook for the chemical industry in the 1970s. He pointed out that the United States economy (which usually reflects that of most of the western community) will look for a 4–4·25% annual increase in its economy throughout the 1970s. However, for the chemicals business, in the period from 1968 to 1980, he prognosticates a two-fold increase in the business equivalent to a 7·5% per annum growth of the business. The organic chemicals business, he expects will increase faster than the inorganics rising at a level of about 9% per annum. In his very well thought out and documented presentation, Dr. Bateman pointed out that there will be a scramble throughout the industry for raw materials to feed the burgeoning appetite of this gigantic organic chemicals business.

The sugar business, through its International Sugar Research Foundation, has no intention of being left in the rear. We search for the information to provide fulfillment of this marketing opportunity.

Sucrose is ubiquitous. Sucrose is available as one of the purest of organic compounds. Its past record in the chemicals business has had no particular distinction, but its properties declare that it has real opportunities. The sugar industry has commissioned us to do what we can to lead the way in demonstrating the utilisation of sucrose by the chemist.

Chapter Seven

SUGAR-MODIFIED MELAMINE RESINS

W. FLAVELL and G. L. REDFEARN

Yarsley Research Laboratories, Chessington, Surrey

Melamine (1, 3, 5-triamino-2, 4, 6-triazine) was first prepared in 1835, and for 100 years remained a laboratory curiosity. Commercial interest was aroused when Henkel in 1935 patented the production of resins by condensation of melamine with formaldehyde. The present production of melamine resins reaches many thousands of tons per annum.

Melamine resins are amongst the most versatile of the thermosetting plastics and are used in the manufacture of moulding materials, laminates, adhesives, surface-coatings, textile-treating resins and paper-treating resins.

The work described herein was sponsored by the Sugar Research Foundation as part of their programme to develop new non-food uses for sugar. The basic concept behind this work was that melamine is an expensive material (at the time when this work was initiated the price of melamine was around 3*s*. per lb), and that by the incorporation of sugar into a melamine-formaldehyde condensate its cost might be reduced without significant loss of properties.

1 CHEMICAL BACKGROUND

Melamine condenses with formaldehyde to give, as initial product, a mixture of water-soluble methylol melamines *(Figure 1)*. The methylol content depends upon the ratio of melamine to formaldehyde and can be as high as hexamethylol melamine.

On further condensation the methylol melamines eliminate water to produce ether links *(Figure 2)*. There is also some evidence that methylene links are produced by reaction between a methylol group and an amino group (eliminating water) or by the elimination of formaldehyde from the methylene ether bridges *(Figure 3)*.

These further condensation reactions are strongly dependent upon the pH and proceed at their slowest rate at around

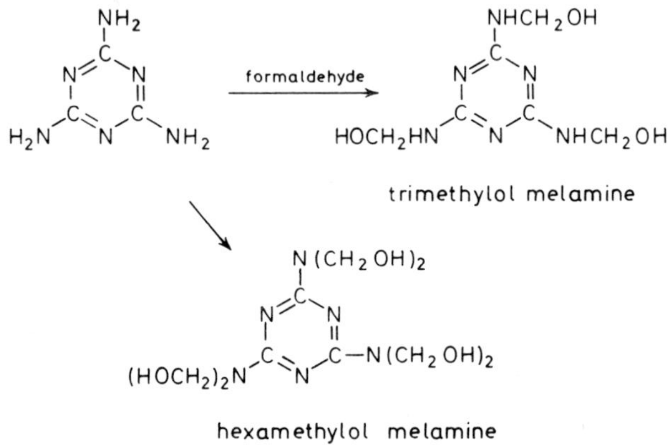

Figure 1. The condensation of formaldehyde with melamine

pH 10. Any increase or decrease in pH gives a marked increase in rate of condensation. This further condensation or resinification leads eventually to the point where the condensate

$$-NHCH_2OH + HOCH_2NH- \;\longrightarrow\; -NH-CH_2-O-CH_2-NH- + H_2O$$

Figure 2. Formation of methylene ether bridges

$$-NH-CH_2OH + H_2N- \longrightarrow -NH-CH_2-NH- + H_2O$$

$$-NH-CH_2-O-CH_2-NH- \longrightarrow -NH-CH_2-NH- + CH_2O$$

Figure 3. Formation of methylene bridges

becomes insoluble in water. In practice the condensation is stopped just short of this point and the aqueous resin solution so obtained is used to impregnate a filler for use in a moulding compound, or a paper substrate for use in a laminate. The final condensation to the insoluble, thermoset or 'cured' state takes place in the fabrication or moulding stage.

It is known that when the melamine-formaldehyde condensation is effected in the presence of alcohols, some reaction takes place between the alcohol and the methylol groups, resulting in ether groups being formed.

$$-NH-CH_2OH + HO-R \longrightarrow -NH-CH_2OR + H_2O$$

This reaction is used to advantage in the preparation of melamine-based surface coatings, which are usually made by condensation of melamine and formaldehyde in the presence of butanol. By this means melamine-formaldehyde resins are obtained which are in solution in butanol and can be blended with other types of resin in other organic solvents.

The knowledge of this etherification reaction suggested the possibility that carbohydrates, and sugar in particular, might be made to combine chemically with the melamine-formaldehyde condensate. In our work on sugar-modified melamine resins for surface coatings, some indirect evidence was obtained that such chemical combination does in fact take place, since the sugar slowly passed into solution in the butanol reaction mixture, and remained in solution. Further indirect evidence on this point was the failure to extract sugar from cured laminates or moulding materials, even under quite vigorous conditions of aqueous extraction.

Preliminary experiments showed that, under conditions approximating to those of resin formation, there is no reaction between invert sugar and melamine, or between invert sugar and aqueous formaldehyde. This lends support to the view that reaction is between the sugar and the initial melamine-formaldehyde condensate and is probably an etherification analogous to that known to occur with butanol under similar circumstances.

Our terms of reference, however, precluded a detailed study of the chemistry of the sugar-modified melamine resins. The bulk of our work was directed towards an empirical investigation of the effects of adding sugar to melamine resins upon the properties of the final products. We have investigated sugar-modified melamine resins in moulding materials, laminates, surface coatings and textile-treating resins. In the last case, however, the results were conflicting and inconclusive, and they are not described in the present paper.

2 MOULDING MATERIALS

2.1 METHODS OF PREPARATION

The manufacture of a melamine moulding material involves the following steps:

1. Preparation of water-based melamine-formaldehyde condensate.
2. Impregnation of filler (usually alpha-cellulose) with this condensate.
3. Drying the impregnated filler.
4. Grinding and incorporation of additives (colourants, lubricants, etc.).
5. Densification, to convert the fine powder obtained by grinding into a form suitable for handling in the moulding factory.
6. Moulding under heat and pressure, in compression presses.

Most of our work on moulding materials was carried out with invert sugar as the modifying carbohydrate, but some work was also done with sucrose, glucose, fructose and corn-syrup.

A typical resin preparation was carried out as follows:

Melamine	504 g, 4 mols
Formaldehyde, 37% w/w	730 g, 9 mols
Invert sugar, 80%	126 g
Aq. ammonia solution, s.g. 0·91	37 ml

The melamine, formaldehyde and invert sugar were charged to a 2-litre flask fitted with stirrer, thermometer and condenser and the pH adjusted to about 8·2 by addition of ammonia. The temperature was raised to 90°C by immersion in a water bath, and the progress of the reaction was followed by removing samples at 1 min intervals and determining water tolerance. On reaching the desired water tolerance (after about 15 min), the resin was neutralised by the addition of dilute sulphuric acid and rapidly cooled to 25–30°C.

The resin so obtained was used to impregnate purified alpha-cellulose using a 1-gallon Z-blade mixer. A normal mixer charge comprised about 600 g resin solution to 270 g cellulose, and mixing was continued for about 30 min. The impregnated cellulose was dried at 70°C in a steam-heated tray dryer, to a moisture content of about 4%, and the dried product was reduced to a fine powder by ball-milling. For our purposes, the densification stage was unimportant and was omitted.

The effects of variations in the following major factors were studied.

1. Ratio of formaldehyde to melamine.
2. Amount of sugar added.
3. Nature of the sugar.
4. Stage of addition of sugar.
5. Ratio of resin to filler.
6. Moulding conditions.

There is considerable interaction between these factors and none of them can be considered in isolation. It is therefore probable that we have not arrived at the optimum combination of factors to give the best results in the final moulded article. However our results indicate the trends resulting from variations in these factors.

TESTS ON MOULDED PRODUCTS

Where appropriate the test methods stipulated in B.S. 1322 'Amino-plastic Moulding Materials' were used. Boiling water absorption is widely used as a control test in the manufacture of aminoplastic moulding materials, and this was adopted as an initial screening test in our work. In addition, the test was extended to include a measurement of soluble matter extracted by the boiling water, since it might have been expected that sugar-modified resins would show a high loss on extraction.

In addition to these screening tests, which were applied to all moulding materials made, the following tests were applied to the more promising formulations.

Mechanical properties:
Tensile strength
Flexural strength and modulus at room temperature
Flexural strength and modulus at 140°C

Electrical properties:
Surface resistivity
Volume resistivity
Electric strength
Resistance to tracking

Miscellaneous properties:
Colour
Colour stability under u.v. light
Resistance to staining.

2.2. RESULTS

Results obtained are summarised in the Tables. *Table 1* shows the effect of varying the invert sugar content and also the effect of varying the moulding temperature upon the boiling water absorption and the loss on extraction. The sugar content is expressed as a percentage by weight of the melamine in the formulation, and the water absorptions and losses as milligrams, based on the use of a moulding of standard dimensions (disc 2 in diameter, $\frac{1}{8}$ in thick) weighing about 10 g.

Table 1. EFFECT OF INVERT SUGAR CONTENT AND MOULDING TEMPERATURE ON WATER ABSORPTION AND EXTRACTION LOSS

Formaldehyde: Melamine = 2·25:1
Resin: Filler = 60:40

% *Invert sugar*	*Boiling water absorption* (mg)		*Extraction loss* (mg)	
	Moulded at 130–135°C	*Moulded at* 142–147°C	*Moulded at* 130–135°C	*Moulded at* 142–147°C
0	96	201	74	76
20	85	110	62	60
30	133	88	67	48
40	138	90	60	60
50	248	100	89	50

The figures in *Table 1* show that, when moulding is carried out at 130–135°C, the addition of invert sugar at a level of more than 20% produces an increase in the boiling water absorption. When moulded at 142–147°C, however, there is little difference between the values given at different sugar contents. The higher figures recorded for the unmodified moulding materials suggests that these might be 'over-cured', and it might be inferred that one effect of the sugar addition is to reduce the rate of curing.

Although moulding at the higher temperature gives rise to lower water absorptions, unfortunately the higher temperature also leads to unacceptably high development of colour when invert sugar is used as modifier. Later work, using sucrose as modifying agent, showed that the higher moulding temperature is essential in order to obtain low water absorptions, and in this case colour development was much lower. These findings came too late in our programme to allow investigation in detail, but some results obtained are given in *Table 2*.

Table 2. EFFECT OF SUCROSE CONTENT AND MOULDING TEMPERATURE ON WATER ABSORPTION AND EXTRACTION LOSS

Formaldehyde: Melamine = 2·25:1
Resin: Filler = 60:40

% sucrose	*Boiling water absorption* (mg)		*Extraction loss* (mg)	
	Moulded at 130–135°C	*Moulded at* 142–147°C	*Moulded at* 130–135°C	*Moulded at* 142–147°C
20	197	83	85	48
30	243	112	54	28
40	215	167	61	45

It will be noted from *Tables 1 and 2* that there is no apparent correlation between sugar content and loss on extraction, again suggesting that the sugar is chemically bonded in the final product.

Table 3. EFFECT OF INVERT SUGAR CONTENT AND MOULDING TEMPERATURE ON TENSILE STRENGTH

Formaldehyde: Melamine = 2·25:1
Resin: Filler = 60:40, *65:35

% *invert sugar*	*Tensile strength (ψ)*	
	Moulded at 130–135°C	*Moulded at* 142–147°C
0	9 800	8 800
20	10 600	12 100
*30	9 000	10 600
*50	8 300	8 100

Table 3 shows data for the tensile strength of modified moulding materials, from which it is apparent that, even at 50% invert sugar (based on melamine) a high tensile strength is obtained.

Some electrical properties are shown in *Table 4*.

Table 4 shows that the electrical properties are impaired by the addition of sugar, but even at 30% sugar content (based

Table 4. ELECTRICAL PROPERTIES OF SUGAR-MODIFIED MELAMINE MOULDING MATERIALS

Formaldehyde: Melamine = 2·25 : 1
Resin: Filler = 60 : 40; *65 : 35
Moulding temperature 130–135°C

% invert sugar	*Surface resistivity* ($\log_{10}$ ohms)	*Volume resistivity* ($\log_{10}$ ohm-cm)	*Electric strength* (volts/mil) (20 s *value at* 90°C)
0	13·5	13·0	175
20	13·0	12·9	135
*30	11·7	12·3	98

on melamine) the electrical properties are still above the minimum requirements of B.S. 1322 Type M.

3. DECORATIVE LAMINATES

3.1 EXPERIMENTAL

The melamine resins used in the manufacture of decorative laminates are basically similar to those used in moulding materials. A water-soluble resin syrup is prepared, and this is used to impregnate paper, which is then dried, several layers stacked together and pressed in a hot press.

The production of satisfactory laminates depends upon several interacting factors and one of the most important of these is the drying or 'pre-cure' stage. If the cure of the resin is too far advanced in this operation, the plies of impregnated paper do not bond to one another during subsequent pressing. On the other hand, if cure is not taken far enough, on hot-pressing too much resin exudes from the laminate and again an unsatisfactory product results.

The bulk of our work on decorative laminates was carried out using one basic resin which was employed either in the unmodified state or modified by the addition of sucrose after preparation. The basic resin formulation was as follows:

Melamine	1 134 g, 9 mols
Formaldehyde solution (37%)	2 187 g, 27 mols

NaOH solution (32%)	ea. 1 ml
Water	165 g
Industrial alcohol	165 g
Borax	3·7 g

The formaldehyde was adjusted to pH 8·4 before adding the melamine, and the mixture was heated to 90°C with stirring and maintained at 90°C until a sample, after cooling to room temperature, became cloudy on dilution with four times its volume of water. At this stage the borax, dissolved in the water and alcohol, was added and the resin cooled rapidly. Where sugar-modified resins were used, sucrose in the desired proportion was added to the pre-formed resin and dissolved by stirring.

Base paper of substance 165 g per sq. m and overlay paper of substance 35–43 g per sq. m were used for impregnation, conditions being adjusted so that the resin contents of impregnated papers were similar when using modified or unmodified resins. Drying and precuring were effected at 100°C in a ventilated oven, varying the time of heating to establish the optimum precure time. Laminates were prepared by pressing two sheets of impregnated base paper and one of overlay paper on to hardboard, at 140°C under a pressure of 44 kg per sq. cm. Pressing times of 8, 11 and 14 min were used.

When using the unmodified resin, laminates of uniform, white glossy appearance were obtained with precure time of 11 min and pressing time of 8, 11 or 14 min. With precure times either longer (14 min) or shorter (8 min), laminates of patchy, dull appearance were obtained. By contrast, resins containing 20% of sucrose (based on melamine) gave satisfactory laminates at all precure times (8, 11 or 14 min) and all pressing times (8, 11 or 14 min). This indicates that the sugar-modified resins are less sensitive to curing conditions and would be expected to have greater latitude in use, leading to a reduction in rejects.

A further advantage was found when the scratch resistances of the laminates prepared from modified and unmodified resins were compared, when the modified laminates were found to be appreciably more scratch resistant *(Table 5).*

The resistance to staining of modified and unmodified laminates was compared, using a variety of staining agents (dyestuffs, hot oil, lipstick, mustard, jam, coffee, tea). No significant differences were found, none of the laminates being stained to any appreciable extent.

The effect of increasing the sucrose content to 50% of the melamine was investigated and similar results obtained, the resins again showing greater latitude in precure and curing conditions, coupled with scratch resistance and stain resistance in no way inferior to the unmodified resins. Long-term storage tests at 25°C were carried out on the unmodified resin and the resin containing 50% sucrose. The viscosity of the unmodified resin increased rapidly and after 72 h was too great to be measured, while that of the modified resin increased at a much slower rate, the resin gelling only after 184 h.

Table 5. SCRATCH RESISTANCE OF LAMINATES

Sucrose content (% on melamine)	*Precure time*	*Pressing time*	*Scratch hardness no.*
0	11	8	16
0	11	11	17
0	11	14	25
20	8	8	26
20	8	11	26
20	8	14	46

Immersion of the laminates (modified or unmodified) in boiling water for 30 min had no effect on gloss or general appearance.

Further increase in sucrose content to 65% of the melamine gave resins which were too viscous to be acceptable in commercial practice, as the rate of impregnation of the paper was too slow. However, laminates were prepared which were of uniform white appearance with a good uniform gloss, using precure time of 8 min and pressing times of 11 or 14 min. Longer precure times gave laminates with slightly less gloss.

3.2 ECONOMICS OF SUGAR MODIFICATION

The economic factors involved in sugar modification are dependent on the prices of sucrose, melamine, formaldehyde and paper, and can thus vary from country to country and from time to time. In the U.K., at the materials costs prevailing at the time of this work, modification of the resin by adding 50% of sucrose, based on the melamine, would result in a reduction in materials cost of 8–10% for the surface layers of the decorative laminates.

ACKNOWLEDGEMENT

This work was financed in the form of a research grant by The International Sugar Research Foundation, Inc.

Chapter Eight

THE CHEMICAL SYNTHESIS OF SUCROSE PHOSPHATES

J. G. BUCHANAN

*Department of Organic Chemistry, University of Newcastle upon Tyne**

There are two reasons for recent interest in the synthesis of sucrose phosphates. The first is that some synthetic preparations have shown an inhibitory effect on dental caries. The second is the possibility of synthesis of the naturally occurring sucrose 6′-phosphate, an intermediate in one of the biosynthetic pathways to sucrose.

A major problem in the synthesis of sucrose derivatives of known structure is the presence of eight hydroxyl groups in the sucrose molecule (I). In 1910 Neuberg and Pollak[1] prepared a sucrose monophosphate as the calcium salt, by treatment of an aqueous solution of sucrose, in the presence of calcium oxide, with phosphorus oxychloride in chloroform. The structure of the product was examined by Hatano[2] and later by Courtois and Ramet.[3–6] Hatano showed that the sucrose phosphate could be hydrolysed with acid to yield D-fructose and a mixture of D-glucose phosphates, 'Hatano's ester'. Courtois and Ramet purified this phosphate mixture by recrystallisation of quinine salts. They concluded, from oxidation studies with periodate and with bromine water, that the major component of Hatano's ester was D-glucose 2-phosphate (II). To our knowledge, no recent work has been published on the structure of Neuberg's sucrose phosphate, but it is very likely to be a complex mixture of isomers, as suggested by Courtois.[6a]

Modifications in the Neuberg synthesis have recently been reported in patents issued to the Colonial Sugar and Refining

*Present address: Chemistry Department, Heriot-Watt University, Edinburgh.

Co. Ltd. of Australia.[7,8] The sucrose monophosphate, as a calcium salt, is an inhibitor of dental caries.[9–11a]

Our own work has been directed towards the synthesis of sucrose phosphates of known structure, in particular the naturally occurring sucrose 6′-phosphate.[12] It has been

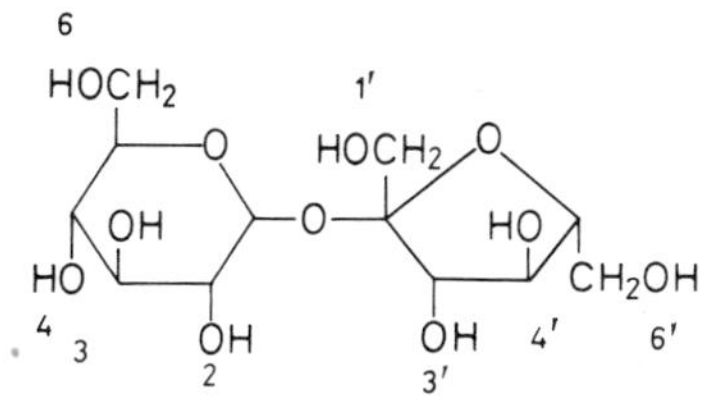

I

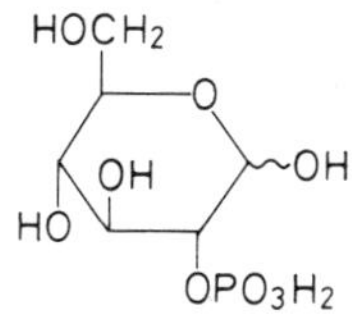

II

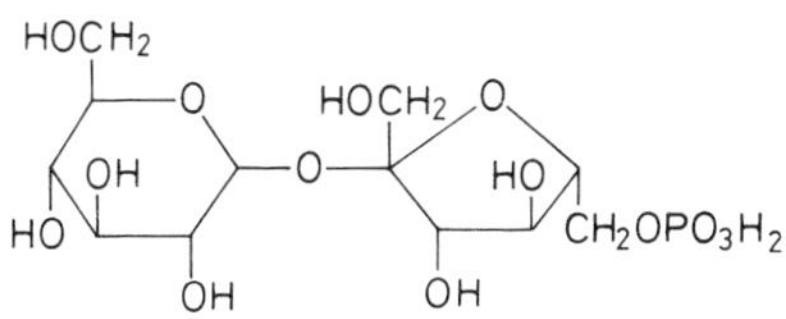

III

carried out in collaboration with Dr. D. A. Cummerson and Dr. D. M. Turner.

The natural occurrence of a sucrose monophosphate in sugar beet leaves was first shown by Buchanan,[13, 14] working in Calvin's laboratory in 1952, fulfilling a prediction made by Calvin and Benson in 1949.[15] Its role in sucrose biosynthesis,[15] being formed by way of nucleotide co-enzyme intermediates, was postulated.[13, 14] The phosphate group was located on the fructose moiety, but the precise location was not proved. Tentatively, the phosphate group was assigned to the 1′-hydroxyl group of the fructose part, but various groups of workers have since shown that the phosphate group in samples of sucrose phosphate from several plant sources is located on the 6′ position, as in III.[12, 16, 17]

Because of its size, triphenylmethyl chloride (trityl chloride, IV) is a reagent with a preference for the less hindered primary hydroxyl groups. By treatment of sucrose (I) with trityl chloride in pyridine, followed by acetylation with acetic anhydride, McKeown, Serenius and Hayward[18] prepared the fully protected derivative V. When trityl groups were removed with aqueous acetic acid the pentaacetate VII resulted, the intermediate pentaacetate VI rearranging to VII during the hydrolysis.[18, 19] In VII there are now only two primary hydroxyl groups, 1′ and 6′. Careful tritylation of VII yielded a monotrityl ether VIII which was not isolated, but treated with alkali to give the crystalline trityl derivative

$(C_6H_5)_3CCl \equiv TrCl$

IV

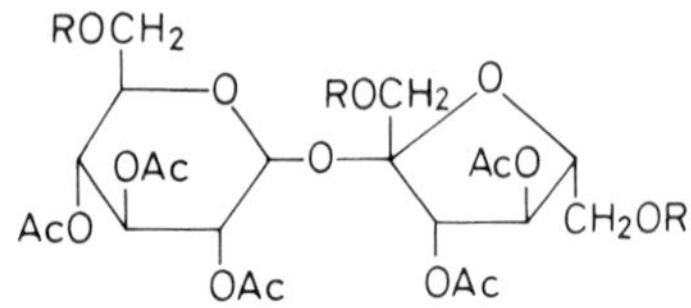

V , R = Tr

VI , R = H

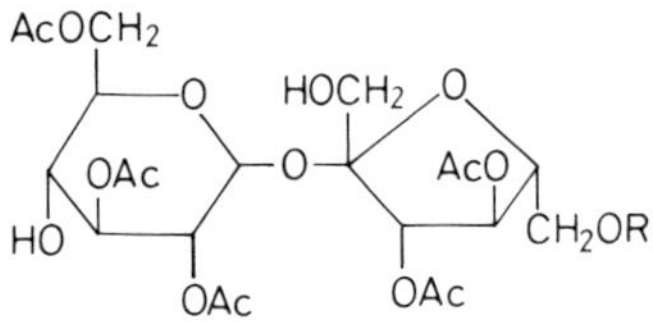

VII , R = H

VIII , R = Tr

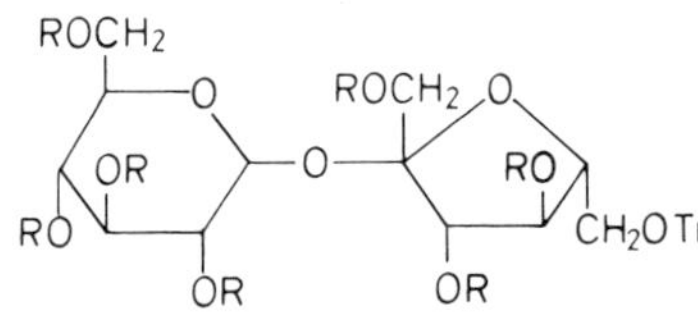

IX , R = H

X , R = Me

XIII , R = Ac

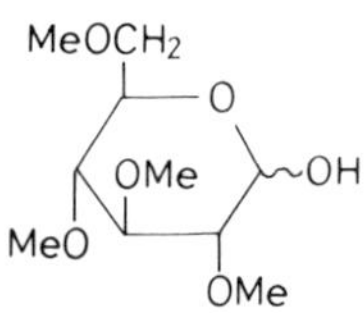

XI

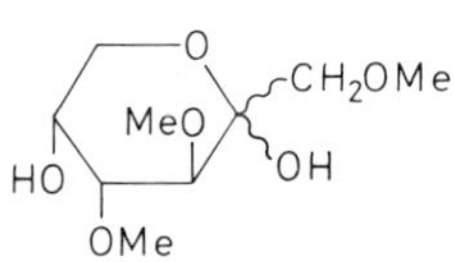

XII

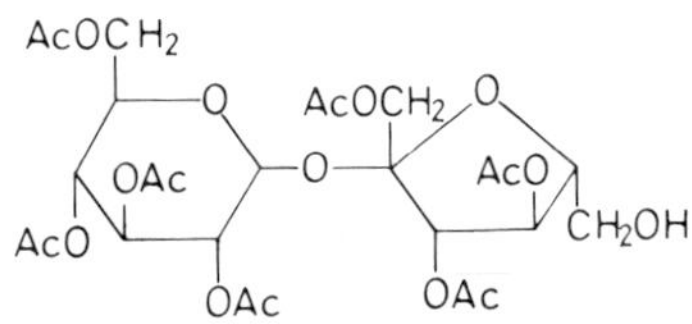

XIV

IX, m.p. 221–223°, in 53% yield from VII. The structure of IX was proved by methylation, to yield the heptamethyl ether X, which gave the two known methyl ethers 2, 3, 4, 6-tetra-*O*-methyl-D-glucose (XI) and 1, 3, 4-tri-*O*-methyl-D-fructose (XII).

We had now selectively differentiated the 6′-hydroxyl group in sucrose. Acetylation of IX gave a crystalline hepta-acetate XIII, m.p. 118°, which was detritylated with aqueous acetic acid to give the crystalline sucrose heptaacetate XIV, m.p. 160°, in 83% yield from IX.

The heptaacetate XIV was phosphorylated by Tener's method,[20] using cyanoethyl phosphate (XV) and dicyclohexylcarbodiimide (XVI) in pyridine. Protecting groups were

$NC{\cdot}CH_2{\cdot}CH_2OPO_3H_2$ (XV) $\qquad$ $C_6H_{11}N{=}C{=}NC_6H_{11}$ (XVI)

removed from the crude product by means of barium hydroxide. The sucrose 6′-phosphate (III) was purified by reprecipitation of the barium salt, and then converted into the potassium salt by means of a cation-exchange resin.

Dr. A. J. Keys, of Imperial College, very kindly determined the enzymic activity of the synthetic compound in the wheat-germ sucrose phosphate synthetase system:[12]

$$\text{Sucrose 6}'\text{-phosphate} + \text{Uridine diphosphate} \rightleftharpoons \text{Uridine diphosphate glucose} + \text{Fructose 6-phosphate}$$

When the synthetic compound III, uridine diphosphate and radioactive fructose 6-phosphate were added to the enzyme, thin layer chromatography clearly showed incorporation of radioactivity into the sucrose phosphate. No reaction was observed in the absence of any of the components.

We thank Dr. Keys for his enzymic determinations, and Dr. D. J. Bell for a sample of 1, 3, 4-tri-*O*-methyl-D-fructose.

REFERENCES

1. NEUBERG, C. and POLLAK, H., *Biochem. Z.*, **23**, 515 (1910)
2. HATANO, J., *Biochem. Z.*, **159**, 175 (1925)
3. COURTOIS, J. and RAMET, M., *Annls Ferment.*, **8**, 105 (1943)
4. COURTOIS, J. and RAMET, M., *C. r. hebd. Séanc, Acad. Sci., Paris,* **218**, 360 (1944)
5. COURTOIS, J. and RAMET, M., *Bull. Soc. Chim. biol.*, **27**, 610, 614 (1945)
6. COURTOIS, J., *Enzymologia,* **11**, 313 (1945)
6a. BRADY, B. H. G., NAPPER, D. H. and SMYTHE, B. M., *J. dent. Res.*, **47**, 603 (1968)
7. French Patent 1,351,134 (1964); *Chem. Abstr.*, **60**, 13426 (1964)
8. British Patent 983,340 (1965); *Chem. Abstr.*, **62**, 14807 (1965)
9. Belgian Patent 638,645 (1964); *Chem. Abstr.*, **63**, 1670 (1965)
10. British Patent 1,009,957 (1965); *Chem. Abstr.*, **64**, 1905 (1966)
11. LILIENTHAL, B., BUSH, E., BUCKMASTER, M., GREGORY, G., GAGOLSKI, J., SMYTHE, B. M., CURTIN, J. H. and NAPPER, D. H., *Aust. dent. J.*, **11**, 388 (1966)
11a. BEVERIDGE, J., GREGORY, G., HARRIS, R., SCHAMSCHULA, R. G. and SAVAGE, D. C. L., *Med. J. Aust.*, **2**, 256 (1968) and references cited therein
12. LELOIR, L. F. and CARDINI, C. E., *J. biol. Chem.*, **214**, 157 (1955)

13. BUCHANAN, J. G., BASSHAM, J. A., BENSON, A. A., BRADLEY, D. F., CALVIN, M., DAUS, L. L., GOODMAN, M., HAYES, P. M., LYNCH, V. H., NORRIS L. T. and WILSON, A. T., *Phosphorus Metabolism,* Vol. 2, John Hopkins Press, Baltimore (1952)
14. BUCHANAN, J. G., *Archs Biochem. Biophys.,* **44,** 140 (1953)
15. CALVIN, M. and BENSON, A. A., *Science, N.Y.,* **109,** 140 (1949)
16. BURMA, D. P. and MORTIMER, D. C., *Archs Biochem. Biophys.,* **62,** 16 (1956)
17. BEAN, R. C. and HASSID, W. Z., *J. Am. chem. Soc.,* **77,** 5737 (1957)
18. McKEOWN, G. G., SERENIUS, R. S. E. and HAYWARD, L. D., *Can. J. Chem.* **35,** 28 (1957)
19. LEMIEUX, R. U. and BARRETTE, J. P., *J. Am. chem. Soc.* **80,** 2243 (1958)
20. TENER, G. M., *J. Am. chem. Soc.,* **83,** 159 (1961)

Chapter Nine

CHEMICAL PROBLEMS IN THE SUCROSE INDUSTRY

K. J. PARKER
Tate and Lyle Ltd., Research Centre, Keston, Kent

INTRODUCTION

The sucrose industry is primarily concerned with the extraction and purification of sucrose. While there are many problems associated with the physical chemistry of the refining processes, and the chemical nature of the impurities extracted from the cane or beet with the sugar, these are not directly related to the chemistry of sucrose.

Apart from hydrolysis, a reaction to which it is particularly susceptible, sucrose is relatively inert. Hydrolysis, frequently referred to as inversion owing to the reversal of the sign of optical rotation of a sucrose solution on hydrolysis, is catalysed by the hydrogen ion and by certain enzymes, non-specifically termed invertase.

Enzyme catalysed inversion is chiefly encountered during the initial extraction of the juice from cane or beet. It is countered by rapidly raising the temperature of the juice to above 80°C in order to inactivate the enzymes. At this stage the pH of the juice is relatively low (5·2–5·8) and acid catalysed hydrolysis does occur. This is minimised by raising the pH with the addition of lime. The effect of this is to initiate a sequence of reactions leading to the formation of the brown polymer chiefly responsible for the colour of raw sugar. Consequently, hydrolysis not only represents a loss of a valuable commodity, but also contributes appreciably to the cost of refining through colour formation, which takes place throughout the refining process.

SUCROSE HYDROLYSIS

The acid catalysed hydrolysis of sucrose has been widely studied giving rise to a voluminous literature. Most of the work has been undertaken with dilute sucrose solutions in the presence of strong acids, using the change in optical rotation to follow the course of the reaction. The reaction kinetics are expressed in terms of acid concentration which leads to complicated relationships definining hydrogen ion activity, while the dependence of rate on sucrose concentration is neglected. In the sugar industry such conditions are rarely met and for practical purposes no suitable data existed from which to calculate rates of hydrolysis under refinery conditions.

Polarimetric methods of analysis are not suitable for higher sucrose concentrations and elevated temperatures. Over the pH range 2–6, the products of hydrolysis, glucose and fructose, are sufficiently stable to permit the progress of the reaction to be followed accurately by the determination of the change in their concentration over a period of time. By choosing conditions which result in not more than 10% hydrolysis, the rate is essentially constant with time. This permits a mean value to be determined for the first order rate constant from successive analyses of the solution.

The sucrose solution, buffered at the required pH, was maintained to within 0·1% at the required temperature. The pH was determined using a Jena glass-thallium amalgam combination electrode. The E.M.F. was measured to 0·1 mV, using a Radiometer pH meter on expanded scale calibrated against a standard Weston cell. The pH of the solution was constant to within 0·002 unit over the period of the run. The concentration of reducing sugars was determined by a cuprimetric method specially developed to give the required accuracy of 0·1%.

Using this method, the rate of sucrose inversion was measured over a range of concentration and temperature, and also in the presence of the chlorides of potassium, sodium, calcium and magnesium. The experimental results are summarised in *Table 1.* The final column gives the value of the second order rate constant extrapolated to zero pH, and corrected for temperature using the mean value of the activation energy of the reaction. The mean deviation of the values is 0·046, which is within the accuracy of the pH definition, considered to be the source of greatest uncertainty.

Substitution of thermodynamic activity for molarity in the concentration term did not lead to a significantly better fit of the data. The effect of salts on reaction rates is not sufficiently pronounced to be considered significant. Neglecting the

Table 1. THE RATE OF SUCROSE INVERSION

Sucrose concentration (g/100 g soln)	*Temperature* °*C*	*pH*	*Salt* (*concentration*)	$\log K_2^\circ + \frac{5810}{T}$
19·92	84·70	4·180		15·276
18·97	80·10	3·702		15·283
40·71	80·00	3·688		15·320
60·16	80·04	3·715		15·363
49·99	80·00	3·702		15·341
20·28	80·02	3·684		15·265
69·63	80·02	3·764		15·395
69·58	80·01	3·798		15·455
20·14	80·01	4·184		15·286
49·55	29·96	1·084		15·299
44·90	40·01	1·489		15·282
37·19	50·00	1·962		15·269
27·86	60·00	2·475		15·230
17·64	70·00	3·074		15·220
6·83	79·99	3·618		15·229
17·80	30·00	1·423		15·311
69·19	80·00	3·640		15·418
65·10	80·01	3·646		15·440
10·12	25·02	1·000		15·294
50·52	25·00	0·966		15·361
65·44	25·00	0·942		15·261
65·20	40·01	1·524		15·270
64·92	50·00	1·985		15·281
64·77	60·00	2·639		15·320
27·00	40·00	1·411		15·351
9·59	70·00	4·129		15·296
9·64	70·00	4·121		15·290
9·60	50·00	4·063		15·308
9·60	70·00	6·140		15·341
58·35	60·00	2·322	KCl (2 M)	15·424
17·68	60·00	2·486	KCl (2 M)	15·416
18·14	60·00	2·510	KCl (1·5 M)	15·394
18·86	60·00	2·568	KCl (1·0 M)	15·421
19·41	60·00	2·601	KCl (0·5 M)	15·390
18·17	60·00	2·603	KCl (1·5 M)	15·431
18·42	50·00	2·015	$CaCl_2$ (1·0 M)	15·209
17·71	50·00	2·062	$CaCl_2$ (1·5 M)	15·134
19·31	50·00	1·965	NaCl (1·0 M)	15·311
18·62	50·00	2·068	$MgCl_2$ (1·0 M)	15·203
9·29	70·00	3·727	KCl (1·0 M)	15·373

marginal influence of salt concentration on the rate of inversion of sucrose, the results may be summarised in an equation relating the first order rate constant to pH, temperature and the molar concentration of water (C_{H_2O}) and sucrose (C_{Suc}):

$$\log_{10} K_1 = 15{\cdot}300 - \frac{5810}{T} - pH_T + \log_{10}(C_{H_2O} - C_{Suc})$$

the pH and concentration being determined at the temperature (T°K) obtaining.

It, therefore, follows that over the experimental range studied, hydrolysis rate is independent of physical variables other than pH and temperature. The free energy of activation is also independent of temperature, pH, sucrose concentration, and dielectric constant and viscosity of the solution. Hydrolysis rate is independent of changes in the activity coefficients of sucrose and water, consequently the rate of the reaction would not be expected to be subject to change in ionic strength, as is found.

Under alkaline conditions sucrose undergoes generalised decomposition with the ultimate formation of acid products, several of which have not been characterised. It is possible to show that neither glucose nor fructose are formed even transiently during the hydroxyl ion catalysed decomposition of sucrose. If the reaction takes place in the presence of the borohydride ion, no significant concentration of the expected hexitol can be detected. Similarly the reducing properties of the solution towards ferricyanide are unchanged in the presence of cyanide ion, whereas if glucose or fructose were present the reducing power would be enhanced.

In the absence of oxygen, one mole of sucrose gives rise to 2·5 equivalents of acid at pH 10·5 at 80°. In the presence of atmospheric oxygen, direct oxidation gives rise to a greatly enhanced acid production.

Consequently, under alkaline conditions the rate of sucrose decomposition must be followed by the direct determination of sucrose. In addition, constant pH must be maintained by the continuous addition of alkali. Uncertainty as to the true value of the pH in sucrose solution at elevated temperature is considered to be the source of greatest error in such a determination. Nevertheless, some results have been obtained which show that the rate of decomposition is directly proportional to hydroxyl ion activity *(Figure 1)*.

This suggests that decomposition of the sucrose anion by a first order reaction probably involves an internal nucleophilic displacement at the glycosidic linkage. This will give rise to an internal anhydride and a reducing sugar moiety in the enolate form. The anhydride will open by the reverse hydroxyl ion displacement, again forming the enolate anion. The further decomposition is extremely complex and depends on the conditions.

It is of interest that the pH at which direct decomposition of sucrose in solution is minimal is around 8·4, approximately

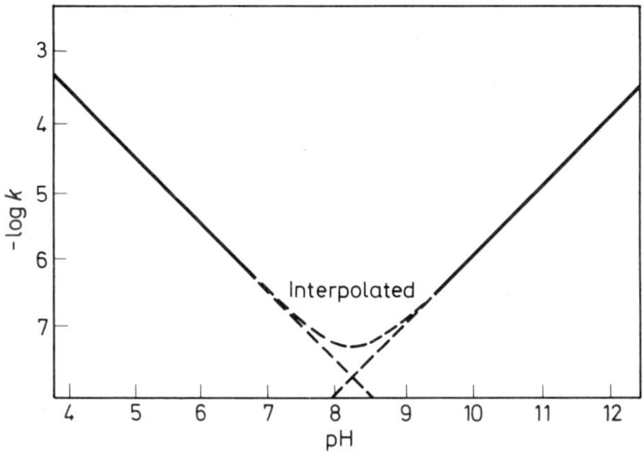

Figure 1. Rate constant of sucrose hydrolysis as a function of pH *at* 80°

two pH units above neutrality. At 80° and at pH 11·30 the rate of decomposition was determined as 2·67 mole^{-1} min^{-1} ≡ 0·47% h^{-1}.

COLOUR FORMATION

The tendency for sugars to form colour on heating—to caramelise—is well known. Apart from this generalised decomposition, colour formation will take place under relatively mild conditions in aqueous solution, though non-reducing sugars are more stable than reducing sugars. The problem of colour formation during sugar production and refining would not exist if inversion of sucrose could be avoided.

Glucose and fructose are relatively stable in aqueous solution at temperatures up to 100° over the pH range 2–7.[1] Under mildly alkaline conditions, especially in the presence of primary amines, decomposition readily occurs with the production of coloured polymers, the optimal pH lying around 9·5. The mechanism of the reaction has not been elucidated, though several routes are probably involved.

On heating an alkaline solution of glucose or fructose, the formation of colour is preceded by the development of a strong absorption in the ultra-violet, with a maximum at 265 nm. The intensity of this peak rises to a maximum and then remains constant, while the absorption gradually extends into the visible spectrum. On stopping the reaction the intensity at 265 nm gradually decays. The absorption appeared to be associated with the formation of an unstable intermediate in the colour forming reaction.

The intermediate responsible was isolated[2] by freezing the equilibrium, selectively concentrating it on ion-exchange resins, followed by steam distillation, further concentration and solvent extraction. It was obtained as a volatile white crystalline material with a characteristic odour. It was too unstable to obtain a complete structural analysis. However, the ultra-violet and infra-red spectra are consistent with the structure (I) assigned to the volatile component.

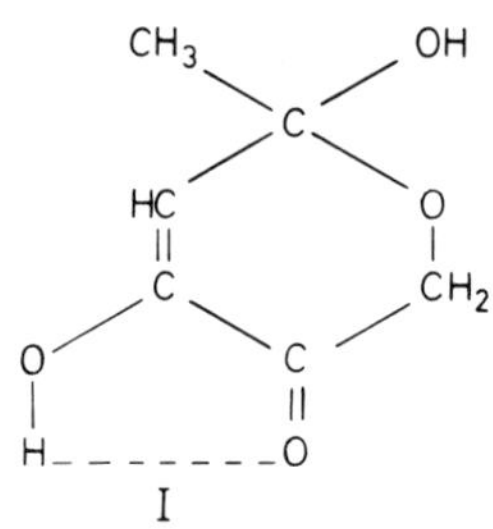

I

$\lambda_{max} = 255 nm$

The mass spectrum of a component separated from the partly decomposed material by gas liquid chromatography of the TMS derivative also supports this structure. A second component in equilibrium with I has λ_{max} = 272 nm. This has not been purified.

The intermediate readily forms a characteristic brown polymer similar in all respects to that isolated from raw sugar.

The coloured polymer has been fractionated, but whatever separation is applied the infra-red spectrum remains unaltered, though the visible spectrum may differ in a characteristic manner. The visible spectrum of all colorants[3] of this type isolated from whatever source conform to the equation relating the absorbance of the solution (A) to wavelength (λ):

$$A \,.\, \lambda^n = \text{constant}$$

The wavelength index (n) may vary, having values between 4 and 10. The coloured polymer is thus remarkably reproducible, suggesting a simple repeating unit. Since the polymer has carboxylic acid, hydroxyl and carbonyl groups in addition to conjugated ethylenic bonds, there being no detectable aromatic protons, its structure is consistent with its being a random polymer of an intermediate such as the dideoxyhexotriulose(I). Polymerisation by successive aldol condensation, would be followed by loss of water and benzyilic acid type rearrangement of the α-dicarbonyl groups. The progressive elaboration of carboxyl groups by the polymer with time has been followed experimentally.

The mechanism of formation of the postulated intermediate must be consistent with the kinetics of the reaction. Cogent evidence has been obtained in support of the following route. The reactions leading to the formation of 3-deoxyglucosulose from glucose by β-hydroxyl elimination are well established.[4] Further hydroxyl elimination gives rise to 3:4-dideoxyglucosulos-3-ene, or by rearrangement to metasaccharinic acid. The analogous β-hydroxyl elimination of 1-deoxyhexo-2:5-diulose will give the intermediate I. However, 3-deoxyglucosulose can also form colour, and may do so by a reverse aldol condensation, followed by recondensation. The rate of colour formation is catalysed by calcium ions, though the yield is reduced, presumably owing to competitive catalysis of the saccharinic acid rearrangement.

The Maillard reaction[5] probably involves the more favourable elimination of the glycosylimine group from the diglycosylamino-acid intermediate, leading to the same deoxyglycosulose, as proposed by Anet.[6] It has been shown that the same coloured polymer results from the Maillard reaction. The enhanced colour obtained is due to a higher conversion of the reducing sugar to colorant.

Although the amine is acting effectively as a catalyst, it does become incorporated into the coloured polymer, which radically affects its physico-chemical properties. Side reactions also lead to the formation of nitrogen ring compounds many of which are fluorescent. These are only lightly coloured. They have not so far been characterised.

Consequently, the colorant is found to differ widely in its solution properties despite its common origin. It is found to range from highly polar to non-ionic, anionic to amphoteric and cationic, hydrophilic to water-insoluble and in molecular weight from a few hundred up to around 10^5. This variation aggravates the problems of decolorisation in sugar refining. However, a knowledge of the nature of the problem permits a rational approach to its solution.

REFERENCES

1. FLEMING, M., PARKER, K. J. and WILLIAMS, J. C., *Proc. 13th Congr. int. Soc. Sug. Cane Technol.,* Taiwan, 1781 (1968)
2. PARKER, K. J. and WILLIAMS, J. C., *Proc. 1968 Tech. Session on Cane Sugar Refining Res.*. 117 (1968)
3. PARKER, K. J., *Proc. 1966 Tech. Session on Cane Sugar Refining Res.,* 72 (1966)
4. ANET, E. F. L. J., *Adv. Carbohyd. Chem.,* **19,** 181 (1964)
5. ELLIS, G. P., *Adv. Carbohyd. Chem.,* **14,** 63 (1959)
6. ANET, E. F. L. J., *Aust. J. Chem.,* **13,** 396 (1960); **15,** 503 (1962)

III

SUCROSE IN THE PLANT

Chapter Ten

THE ROLE OF SUCROSE IN GREEN PLANTS

JACK EDELMAN

Professor of Botany, Queen Elizabeth College, University of London

Sucrose is found without exception in all the green plants we see growing around us. It occurs as an early product of photosynthesis, as the main agent for translocating carbon from the photosynthetic centres to the rest of the plant, and it often acts as an important reserve substance—in some cases, such as sugar cane, it is the sole carbon reserve found in the plant. Sucrose is important in all these aspects, and this importance is emphasised when we realise that most of the carbon compounds found in the non-photosynthetic parts of the plant, that is, the non-green parts such as roots, internal tissues of stems, flowers and seeds, are derived from it. Sucrose, however, is an unexpected sugar to find as it is something of a chemical curiosity. It is only common in its ubiquitousness but on reflection that is perhaps its most remarkable property of all. We can say, almost without exception as far as we know, that green plants are addicted to sucrose! So in considering the role of sucrose in plants the nub of the problem is 'why sucrose?'. Why for instance does glucose not suffice in plants as it does in many animals including man? Presumably we must discover physical or chemical properties peculiar to sucrose in order to understand why it is so especially important in the metabolism of plants.

SOME CHEMICAL CONSIDERATIONS

First, there are four possible sucroses based upon D-glucose and D-fructose (this number rises to sixteen if we consider all possible combinations of the D and L monomers). But only one of these (I), with D-glucose linked α and D-fructose β, is the sucrose that we know; $\alpha\alpha'$, $\beta\alpha'$ and $\beta\beta'$ sucroses are unknown in nature, and as far as I know the physiological behaviour in plant tissues of artificially synthesised isomeric sucroses has not been investigated. This is potentially a rich

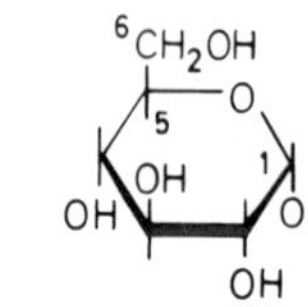

α-D-glucose

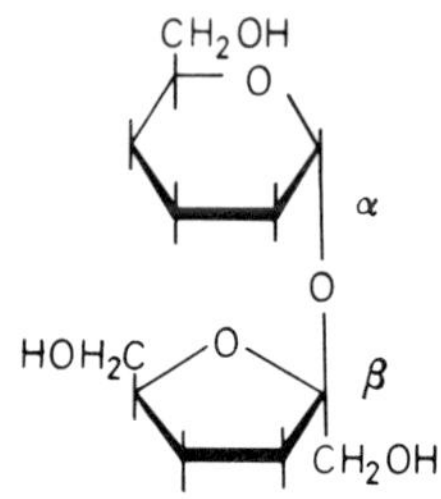

I Sucrose

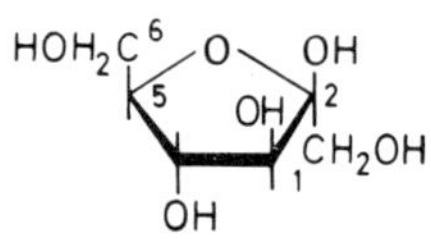

β-D-fructose (furanose form)

field of research, as the behaviour of such compounds may pinpoint which parts of the molecule have the required configuration.

Sucrose is an α-glucoside, but as many other α-glucosides are found in plants, the disaccharide maltose ($\alpha\alpha'$-glucosido-glucose, II) being the best known, we can speculate that it

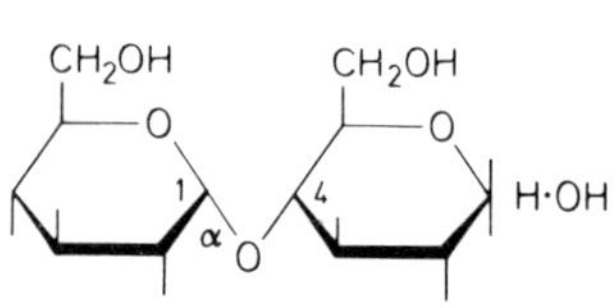

II Maltose

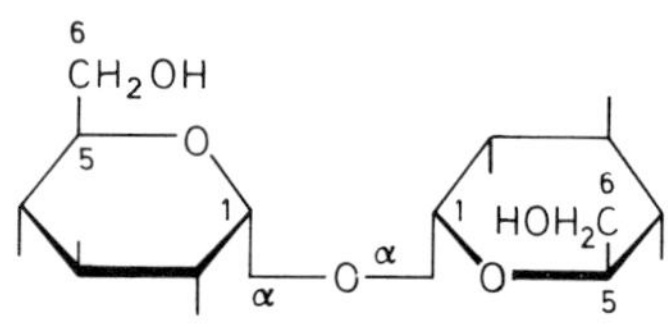

III αα′-Trehalose

cannot be this part of the molecule which confers its essential properties. But sucrose unlike maltose is also a β-fructoside, and it is therefore a double glycoside, i.e. the reducing groups of both of its monosaccharides are joined together, so perhaps it is this rather rare aspect of its structure which is important. This may well be so, as the only other disaccharide of this sort which occurs in nature is trehalose, which is $\alpha\alpha'$-diglucoside (III), and it is interesting that this sugar is found in place of sucrose in the cell sap of some fungi; it also occurs in the circulatory system of insects.

The β-fructoside nature of sucrose is again something of a rarity. It does occur in other compounds but these all appear to be based on the sucrose molecule itself and can be considered derivatives of it. Oligo-saccharides of this sort (e.g.

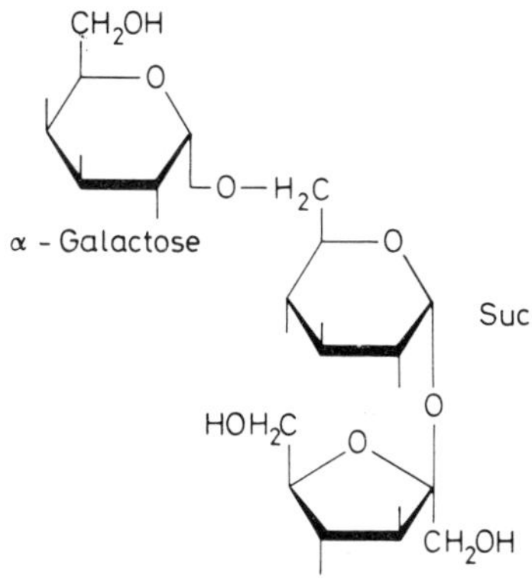

IV Raffinose

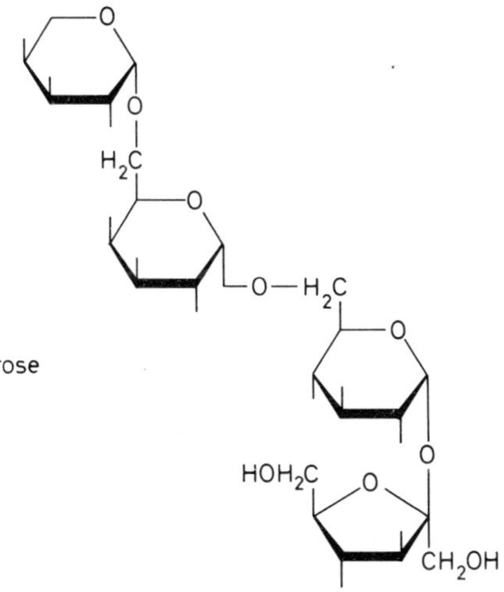

V Stachyose

raffinose IV, stachyose V) are sometimes found in small amounts in the translocatory stream or other places in plants, but it appears likely that it is the sucrose part of such molecules which allows them to be used in this way. Another class of compounds containing β-fructoside linkages is the fructans—the fructose polymers. Various types of fructosan are found in plants, where they are usually considered to be storage carbohydrate. As with the oligomers, they are based upon the sucrose molecule, being fundamentally chains of fructose units terminated by a sucrose grouping (VI, VII). Among flowering plants, fructans are found mainly in the most highly evolved families, namely the Compositae (sunflower family) and the Graminae (grasses and cereals) and the significance of the β-fructoside linkage in these compounds may reflect more than mere storage of carbon, a purpose usually satisfied by accumulation of the α-glucoside polymers amylose and amylopectin which together comprise starch. Certainly fructose is found only in a very restricted number of compounds whereas glucose and many other monosaccharides are found very widely distributed in storage and cell-wall polysaccharides and in other glycosides of various sorts and with a variety of linkages.

The furanose configuration of the fructoside group in sucrose confers upon the linkage between the two moieties a very high free energy of hydrolysis. This is a point which is often overlooked in the consideration of the physiology and biochemistry of sucrose. This available energy content, which

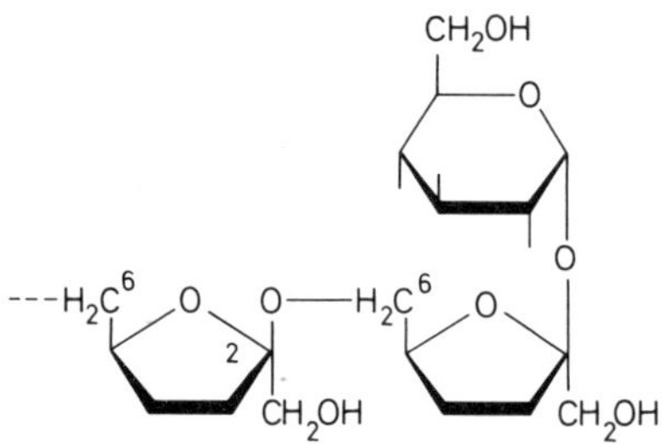

VI Fructan (levan type)

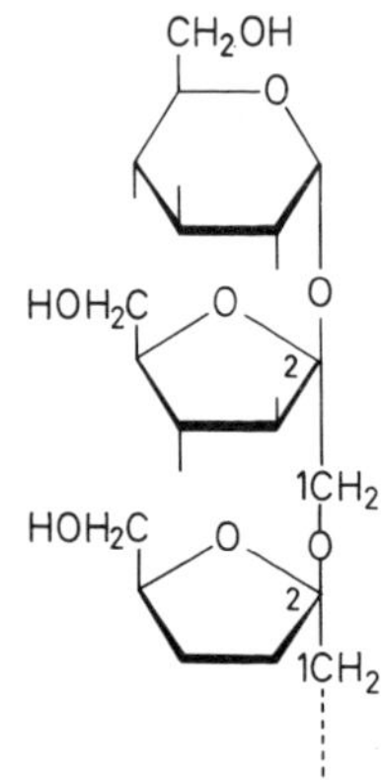

VII Fructan (inulin type)

has been estimated to be as high as −6·6 kcal/mole,[1] is close to that of the 'high energy bond' of ATP, −6·9 kcal/mole. Its significance is entirely unknown.

PHYSICAL CHARACTERISTICS

Apart from osmotic pressure, which is dependent upon molecular weight, various physical characteristics of sucrose are not widely different from those of glucose. Arnold[2] has compared solutions of the two sugars with respect to a variety of their physical characters *(see Table 1)* and although there are some differences between them he concludes that these are insignificant in explaining why sucrose is used for translocation in plants and glucose is not. The level of sugar he used, namely 20%, is comparable to that found in the translocatory stream in the conducting tissues of plants.

Table 1. PHYSICAL PARAMETERS OF WATER AND 20 WEIGHT PER CENT SOLUTIONS OF SUCROSE AND GLUCOSE (from the *Journal of Theoretical Biology*, by courtesy of Arnold[2])

Parameter	*Units*	*Temperature*	*Water*	*Sucrose*	*Glucose*
Density	g/ml	d_4^{20}	0·998	1·081	1·080
Viscosity	millipoises	25	8·94	17·10	16·6
Surface tension	dynes/cm	25	72·0	73·0	73·4
Dielectric constant	$E_{vacuum} = 1$	20	81·9	75·3	55·7
Osmotic pressure	atmos	0	0	17·7	32·0
Diffusion coefficient	$(cm^2/sec)\cdot 10^7$	25	223	39·5	49·0

THE BIOCHEMISTRY OF SUCROSE

If we cannot find any clue to the importance of sucrose as a molecule, despite its special chemical characteristics, we

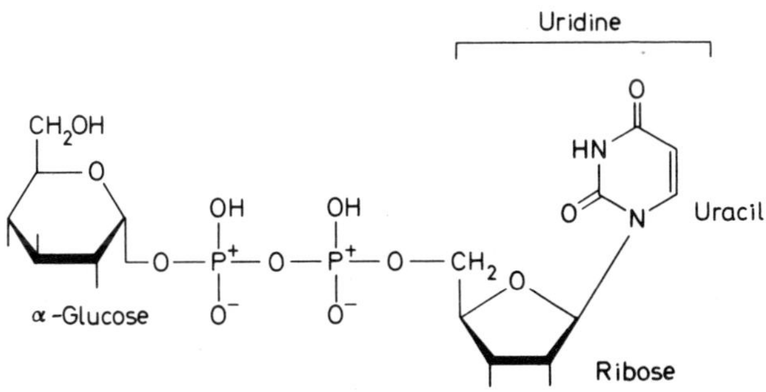

VIII Uridine diphosphate glucose

might be able to find one by examining the biochemical pathways of its synthesis and breakdown. For its synthesis a sugar nucleotide, uridine diphosphate glucose (UDPG, VIII, acts as a glucosylating agent by transferring a glucose residue to fructose-6-phosphate F6P forming sucrose phosphate (G—F6P) under the action of the enzyme sucrose phosphate synthetase:

$$\text{UDPG}+\text{F6P} = \text{G—F6P}+\text{UDP}$$

The sucrose phosphate is then hydrolysed by a phosphatase to sucrose and inorganic phosphate:

$$\text{G—F6P} \rightarrow \text{sucrose} + \text{phosphate}$$

The fructose-6-phosphate is derived from photosynthesis by a series of well-known reactions which we need not go into here (*see* Walker, Chapter 11 in this volume), and the glucose itself in UDPG can also be derived from another photosynthetically produced fructose-6-phosphate molecule in the presence of the nucleotide uridine triphosphate. There is nothing very special about these reactions. Sugar nucleotides are used widely in plant tissues and bacteria for synthesising oligosaccharides and polysaccharides,[3] and sucrose appears to be no exception to this. So there is nothing in this particular sequence which can explain the importance of sucrose in the plant.

What about sucrose breakdown? There are two enzymes known in plants which might be used for this. The first of these is one of the oldest agents known to enzymology—invertase. This enzyme hydrolyses sucrose to free glucose and fructose. Micro-organisms such as yeast which produce the enzyme extracellularly almost certainly use it to convert the disaccharide to the more reactive monosaccharides for further absorption and metabolism. But in plants it is not so obvious that invertase is used for this purpose. In fact it may well be that its role is more complex (*see* below) and may be even chemically different from its action in the test-tube, but we understand no more about this at the moment than we do about the significant parts of the structure of sucrose itself.

The other enzyme which could be involved in sucrose breakdown is sucrose synthetase,[4] named because it was first discovered making sucrose from UDPG and fructose (not fructose-6-phosphate); it is now considered more likely to act in reverse as follows:

$$\text{G} - \text{F} + \text{UDP} = \text{fructose} + \text{UDPG}$$

Neither of these enzyme reactions seem to confer upon sucrose any particular advantage over the free hexoses. It is true that the UDPG formed by sucrose synthetase can give rise to glucose-1-phosphate which is an important metabolite and can be readily absorbed into intermediary metabolism. But this would appear to be of no great advantage as free glucose can equally readily be phosphorylated in the presence of ATP and, in contrast to animals, there is no shortage of ATP in plants (derived either from photosynthesis, or from respiration of the very large carbohydrate reserves which are found in virtually all plant cells). In any case this reaction gives rise to free fructose in equal amounts to UDPG and this would have to be phosphorylated and metabolised as if it were originally free monosaccharide.

In addition to these obvious pathways of synthesis and breakdown a variety of enzymes is now known which will attack sucrose rather differently. These enzymes will transfer either the fructose or the glucose half of the molecule to an acceptor molecule, sucrose thus acting as a glycosylating agent. The acceptor can be another sucrose molecule itself and in this way chains of fructose or glucose units can be added to a sucrose primer, or some other sugar primer, to give polysaccharides, e.g.:

$$\mathrm{GF+GF \xrightarrow{transferase} GFF+G}$$
$$\mathrm{GFF+GF \rightarrow GFFF+G\ etc.}$$

Alternatively it is sometimes found that the fructofuranose moiety can be transferred to a primary alcohol group, e.g. methanol or ethanol, to give the corresponding methyl or ethyl fructosides. Enzymes such as these make the extracellular polysaccharides which are found in some bacteria, but they are also found in plants where their function is more obscure. Certainly the fructose polysaccharides found in plants are built up in this way,[5] but the role of similar enzymes in tissues where polysaccharides of this sort are not found is not yet understood. One of the most interesting aspects of sucrose metabolism has been the discovery that invertase itself, for many years considered to be a merely hydrolytic enzyme, is able to catalyse transfers of this sort to give rise to tri- and tetrasaccharides and even higher oligomers,[6] (*see* equation above) although these are ultimately broken down to the monosaccharides as final products. It may be that the role of invertase in making transitory intermediates of this sort is more important in its metabolic function within the living tissue than its ability to hydrolyse sucrose.

A PROTECTED DERIVATIVE OF GLUCOSE?

The secret of sucrose may lie in the extreme biochemical reactivity of the monosaccharides, glucose and fructose, of which it is composed. Arnold[2] has recently suggested that the disaccharide is a convenient and comparatively unreactive—but not too unreactive—derivative of glucose: that it protects the monosaccharide from metabolic attack during translocation. But why should this be particularly necessary in plants and not in animals? The translocatory material in plants passes at high concentrations (10–20%) through the cytoplasm of each cell of the transporting system (*see* Spanner, Chapter 12 in this volume); but in animals the translocatory sugar is more transitory, it travels around the organism very much more quickly, it exists at a very much lower concentration (0·1% in man) and is extracellular for most of its journey. Perhaps for these reasons the translocatory sugar is not so susceptible to attack.

In plants moreover the sugar may continue to exist in the cell vacuoles both as a storage compound and also as an osmotic agent conferring turgor on the tissue. For both of these roles susceptibility to enzymic attack would be an embarrassment, and a relatively inert structure would be a great advantage. But the problem, why particularly sucrose and not one or a variety of other di- or oligosaccharides remains as yet unanswered.

REFERENCES

1. LELOIR, L. F., CARDINI, C. E. and CABIB, E., *Comparative Biochemistry,* Academic Press, New York (1960)
2. ARNOLD, W. N., 'The Selection of Sucrose as the Translocate of Higher Plants', *J. theor. Biol.,* **21,** 13–20 (1968)
3. HASSID, W. Z., 'Transformation of Sugars in Plants,', *A. Rev. Pl. Physiol.* **18,** 253–280 (1967)
4. CARDINI, C. E., LELOIR, L. F. and CHIIBOGA, J., 'The Biosynthesis of Sucrose', *J. biol. Chem.,* **214,** 149–155 (1955)
5. EDELMAN, J. and JEFFORD, T. G., 'The Mechanism of Fructosan Metabolism in Higher Plants as Exemplified in *Helianthus tuberosus*', *New Phytol.,* **67,** 517–531 (1968)
6. EDELMAN, J., 'The Formation of Oligosaccharides by Enzymic Transglycosylation', *Adv. Enzymol.,* **17,** 189–232 (1956)

Chapter Eleven

THE SITE OF SUCROSE SYNTHESIS IN GREEN PLANTS

D. A. WALKER
Reader in Enzymology, Imperial College, University of London

To anyone accustomed to taking sugar in his morning coffee with pleasure, regardless of the possible consequences, it may come as a surprise to learn that we really know very little about where it is actually made. We know that the green pigment in higher plants is restricted to small discrete organelles called chloroplasts just as the principle red pigment in people is restricted to blood corpuscles (which are not too different in size and shape from chloroplasts). We know that chloroplasts are the site of photosynthesis. That is to say they absorb and utilise energy to convert carbon dioxide and water to carbohydrate and discard oxygen as a waste product.

$$CO_2 + H_2O \rightarrow (CH_2O) + O_2$$

We know that the chloroplasts can fix CO_2 and evolve O_2 as readily as a whole plant because after many years of effort it has proved possible to isolate these organelles from the plant and get them to do just that.[1–5] Moreover, we know that chloroplasts can synthesise starch because the presence of starch grains within chloroplasts can be readily demonstrated whereas no one has been able to catch a chloroplast in the act of engulfing a starch grain, like a man swallowing a beef steak. Until recently, however, it had not proved possible to demonstrate unequivocally that the chloroplast can synthesise sucrose. It is clear, of course, that sucrose can be synthesised at other sites within the plant because it can be produced in tissues which contain no chloroplasts. For example, a single germinating *Ricinus communis* seed can convert castor oil to sucrose at something like 5 mg/h, a feat which we might wish we could emulate. Therefore, although there is no doubt that the primary conversion of CO_2 to triose, pentose and hexose phosphates occurs within the chloroplasts, the possibility that these compounds were exported from the chloroplast to be converted into sucrose elsewhere, could not be excluded. In fact, since no one had been able to get isolated chloroplasts to make sucrose from CO_2 and water, it seemed quite feasible that they might lack the appropriate machinery. Enzyme distribution studies indicated otherwise. As we have heard, sucrose may be synthesised by reactions involving the transfer of glucose from UDPG to fructose-6-phosphate or fructose. If fructose-6-phosphate is the acceptor then the immediate condensation product is sucrose monophosphate.

Work by Bird *et al.*[6] has shown that the enzymes which catalyse these reactions are present in tobacco chloroplasts[7] and that the distribution favours the route which leads to the formation of sucrose phosphate. This is itself consistent with photosynthetic labelling experiments which indicate that sucrose becomes rapidly labelled in both the fructose and glucose parts of the molecule. Since sugar nucleotides + fructose-6-phosphate are also rapidly labelled whereas free fructose accumulates label only slowly, it may be inferred that sucrose phosphate is a precursor of free sucrose. Nevertheless, although enzyme distribution studies are very valuable they must always remain inconclusive while the possibility exists that one fraction may contaminate another during isolation. This is particularly true of nonaqueous extraction where the possibility of precipitating soluble proteins on to organelles during isolation must be very real. Until it proved possible to demonstrate sucrose synthesis by isolated chloroplasts *undergoing photosynthesis*[7] there was therefore an element of doubt. This was finally removed in 1967 when Gibbs and his colleagues[8, 9] at Brandeis University showed convincing evidence of a massive sucrose synthesis by spinach chloroplasts in which more than 60% of the radioactive carbon incorporated into photosynthetic products was distributed equally between the two halves of the sucrose molecule (i.e. in accord with the sucrose phosphate route). The results were entirely unequivocal and reproducible and there, it must have seemed, was the end of the story. However, it was not that easy. What worried Gibbs *et al.* was that like everyone else in the field they had previously obtained endless chromatograms showing no sucrose and, to the best of their knowledge, they had changed nothing in their experimental procedure. What worried them more was that at the end of a few weeks the sucrose departed as abruptly as it had arrived, or to be more precise, the behaviour of the chloroplasts reverted to what had become regarded as the norm and no further sucrose was synthesised. Since this time, sucrose synthesis has again been observed in Gibbs' laboratory and it is possible that this is a seasonal

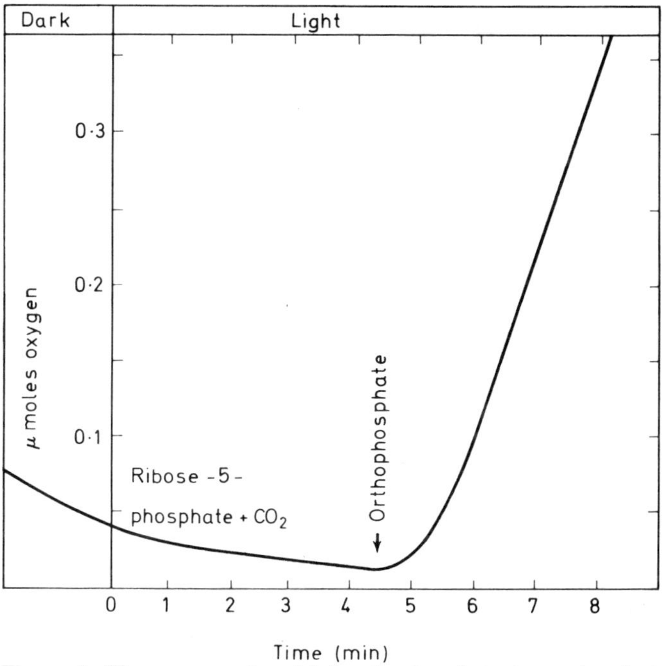

Figure 1. The response to added orthophosphate in a phosphate deficient mixture in which there is thought to be competition between ribulose-5-phosphate kinase and 3-phosphoglycerate kinase for the available ATP

affect. If it is seasonal it is clear that it does not affect the parent tissue in the same way, since spinach leaves make sucrose at all times. It is equally clear that it is not a function of the overall activity of the preparations since sucrose has been synthesised (and not synthesised) by highly active and poorly active chloroplasts alike.

The most probable explanation for this enigmatic situation lies in some delicately poised control mechanism. *Figure 1* for example shows how photosynthesis may be inhibited by an excess of ribose-5-phosphate and how this inhibition may be entirely reversed by the addition of orthophosphate. *Figure 2* shows the converse in which orthophosphate is inhibitory and the inhibition is reversed by the addition of ribose-5-phosphate. These experiments[10] may have no direct bearing

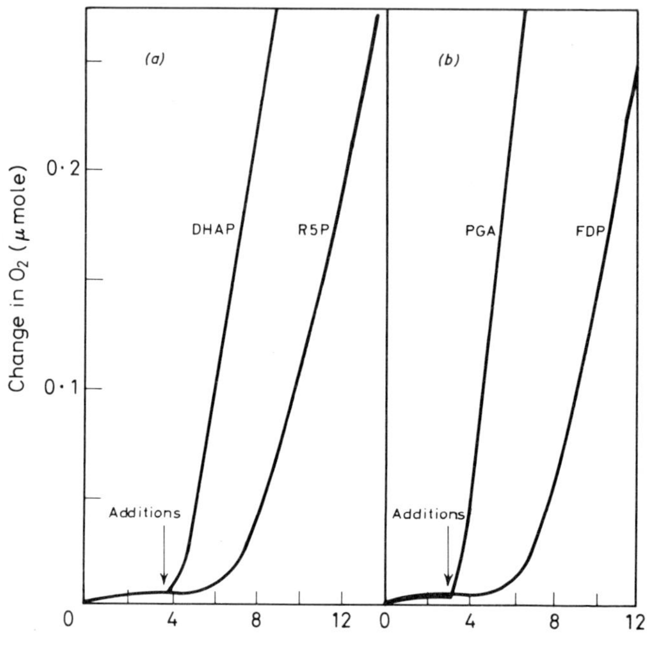

Figure 2. The reversal of orthophosphate inhibition by phosphoglycerate, dihydroxyacetone phosphate, ribose-5-phosphate and fructose-1, 6-diphosphate

on sucrose synthesis; we have no particular reason to suppose that they do, but I have included them here because they illustrate two things. One is that under these conditions both ribose-5-phosphate and orthophosphate can penetrate the chloroplast envelope quite freely; the second, that the rate of photosynthesis can be governed by a balance between orthophosphate and a sugar phosphate. Clearly if photosynthesis can be switched on and off simply by altering the balance between such metabolites it is not improbable that the traffic of carbon into a single component such as sucrose might be

Table 1. SUCROSE SYNTHESIS BY ENZYMES ASSOCIATED WITH ISOLATED CHLOROPLASTS
(Courtesy of C. W. Baldry and I. F. Bird)

	*NAC**[14]	^{14}C–UDPG 1·2mmoles	F6P 2mmoles	*GK20*†	*Gluta-thione*	*R.M.*‡	$NaHCO_3$	*A.C.*§	*Tricine* pH 7·5 7·3mmoles	H_2O	*Sucrose Synthesis*
Vol (ml)	0·05	0	0·02	0·05	0·02	0·10	0·02	0·10	0·05		
1	√	√	√	—	—	—	—	—	√	√	Positive
2	√	√	√	√	√	—	√	√	—	√	Positive
3	√	√	√	√	√	√	—	—	—	√	Positive
4	√	√	√	√	√	—	—	√	—	√	Positive
5	√	√	√	—	—	—	√	—	√	√	Positive
6	√	√	—	√	√	—	√	√	—	√	Negative
7	—	√	—	√	√	—	√	√	—	√	Trace
8	√	√	—	√	√	—	—	√	—	√	Trace

Final volume of digests 0·3 ml
*NAC—Non-aqueous chloroplasts—prepared[6] from freeze-dried tobacco leaves (var. White Barley). 20 mg of dry chloroplasts resuspended in 0·5 ml Tricine buffer pH 7·5 and suspension centrifuged. 0·05 ml of the supernatant, equivalent to 2 mg dry weight of chloroplasts, added to the digests indicated.
†GK20—A solution containing cofactors such as Mg^{++}, Mn^{++}, etc.[1]
‡R.M.—Resuspending medium.[1]
§A.C.—Aqueous chloroplasts prepared, in sorbitol,[1] from peas (Laxton Superb).
In order to keep the volume of the digests small the solution of ^{14}C–UDPG was freeze-dried in the reaction tubes prior to the addition of the other materials.
The reaction tubes were illuminated from two hours with saturating red light.
Temperature 20°C

similarly controlled. It would follow, since these metabolites are able to move through the chloroplast envelope, that sucrose synthesis might be partly controlled from the cytoplasm.

Shortly before he moved from Imperial College to work for Tate & Lyle, my colleague, C. W. Baldry, undertook some joint exploratory experiments with I. F. Bird. I have their permission to reproduce the results of one of these experiments in *Table 1.* It shows that chloroplasts isolated in aqueous media (which were able to fix CO_2 at high rates) contained nothing to inhibit sucrose synthesis and nothing which degraded it once it had been synthesised by the non-aqueous fraction. Had the aqueous chloroplasts only been able to provide fructose-6-phosphate there seems to have been no reason why these mixtures should not have synthesised sucrose from added CO_2. On the other hand fructose-6-phosphate is an intermediate of the Benson-Calvin cycle and we have substantial evidence which supports the view that the aqueous chloroplasts operate this cycle at high rates.[1–5, 10] Moreover, fructose-6-phosphate will apparently pass through the chloroplast envelope[10] but its addition is seemingly without effect on sucrose synthesis by aqueous chloroplasts.

There for the moment the problem must rest. The evidence that chloroplasts possess the sucrose synthesising machinery is compelling. Why they fail to operate it in most circumstances is a mystery. Several possibilities warrant further investigation. One is that the enzymes directly concerned undergo allosteric activation or inhibition.[11] Another is that fructose-1, 6-diphosphatase plays as important a regulatory role[12, 13] in photosynthesis as it does in animal glucogenesis.[14] Our own feeling is that, in addition, sucrose synthesis is unlikely to be independent of the mechanisms governing the passage of metabolites into and out of the chloroplast. If any (or all) of these factors *are* involved the answer to this problem may turn out to be more important than the problem itself.

REFERENCES

1. WALKER, D. A., *Biochem. J.,* **92,** 22–23c (1964)
2. WALKER, D. A. and HILL, R., *Biochim. biophys. Acta.,* **131,** 330–38 (1967)
3. JENSEN, R. G. and BASSHAM, J. A., *Proc. natn. Acad. Sci. U.S.A.,* **56,** 1095–1101 (1966)

4. COCKBURN, W., WALKER, D. A. and BALDRY, C. W., *Pl. Physiol., Lancaster,* **43,** 1415–8 (1968)
5. WALKER, D. A., BALDRY, C. W. and COCKBURN, W., *Pl. Physiol., Lancaster,* **43,** 1419–22 (1968)
6. BIRD, I. F., PORTER, H. D. and STOCKING, C. R., *Biochim. biophys. Acta,* **100,** 366–75 (1965)
7. HAQ, S. and HASSID, W. Z., *Pl. Physiol., Lancaster,* **40,** 591–94 (1965)
8. EVERSON, R. G., COCKBURN, W. and GIBBS, M., *Pl. Physiol., Lancaster,* **42,** 840–44 (1967)
9. GIBBS, M., LATZKO, E., EVERSON, R. G. and COCKBURN, W., *Harvesting the Sun,* Ed. by San Pietro, A., Greer, F. and Army, T. J., Academic Press, New York, 111–130 (1967)
10. COCKBURN, W., WALKER, D. A. and BALDRY, C. W., *Biochem. J.,* **107,** 89–95 (1968)
11. PREISS, J., GHOSH, H. P. and WITTKOP, J., *Biochemistry of Chloroplasts,* Academic Press, New York, **2,** 131–153 (1966)
12. BASSHAM, J. A. and JENSEN, R. G., *Harvesting the Sun,* Ed. by San Pietro, A., Greer, F. A. and Army, T. J., Academic Press, New York, 79–110 (1967)
13. BUCHANAN, B. B., KALBERERER, P. P. and ARNON, A. A., *Biochem. biophys. Res. Commun.,* **29,** 74–79 (1967)
14. KREBS, H. A., *Bull. John Hopkins Hosp.,* **95,** 19–29 (1954)

Chapter Twelve

THE MOVEMENT OF SUCROSE THROUGH PLANT CELLS

D. C. SPANNER
Reader in Botany, Bedford College, University of London

One of the classical problems of plant physiology, handed down from the early days when the Reverend Stephen Hales (1677–1761) corresponded with Malpighi about the results of ringing experiments on trees, was the secret of how substances moved about in plants. Was it by a process analogous to the circulation of the blood in animals? In those days knowledge of plant nutrition was in a very rudimentary state, and it was generally believed, following Aristotle, that the plant gained all its carbonaceous matter from the soil; though after this had been passed up to the leaves it was transformed in various ways under the action of sunlight and moved to other parts of the plant. Building on the foundation of these early workers we now know that the carbonaceous material of the plant body is derived almost exclusively from the air, and exported from the leaves where it is fixed in chemical form to the growing extremities of the roots and shoot. The details of this long-distance transport process constitute the problem of translocation. Although it is a problem which we can now define rather precisely, and which is fairly accessible to the experimental approach, it is still one which remains a very considerable mystery. It is about this problem that I wish to speak, and I shall do so under four heads: the sugar involved, the pathway of movement, the velocity of transport, and the mechanism employed.

THE SUGAR INVOLVED

Sugars being the primary products of photosynthesis the question naturally arises, in what form are the products of assimilation moved in bulk from the leaves to the growing points? The obvious way to investigate this problem would be to extract the sap from the conducting channels, once these have been ascertained, and analyse it. This has now been done for a number of plants using chromatographic methods of analysis, which of course require only a small amount of sap; but it is interesting to remember what sterling work was done before the advent of many of the elegant modern methods of micro-analysis. It seemed a fairly safe guess that sugars would form the bulk of the assimilate stream; and in a splendid series of painstaking researches on cotton in the 1920s Mason and Maskell showed fairly conclusively that sucrose was the form in which sugar was transported in this plant. They did this by following the variations in concentration in the three principal sugars of the leaf and stem, glucose, fructose and sucrose. While the hexoses vary widely in concentration during the course of the day and night in the leaf, they are fairly constant in the stem. However, the sucrose in the bark of the stem follows the ups and downs of the leaf hexoses very closely, with a slight time lag. The presumption therefore is that sucrose represents the form in which the sugar is being transported, and that it is being conveyed in one of the tissues included in the bark.

Another method requiring relatively large amounts of plant material was employed by Tammes. He took advantage of the fact that certain arborescent monocotyledons with large inflorescences exude very considerable quantities of sap if the young inflorescence is cut off. This sap almost certainly comes from the phloem, and it can be collected and analysed in bulk. Using the palm *Arenga saccharifera,* Tammes found that the sap contained about 15% of sucrose but no other sugars. In addition it contained about 0·5% amino acids and about 0·12% potassium. He has more recently used the same method with *Yucca*, and found about the same amount of sucrose and about 0·5% hexoses. Invertase was significantly absent.

Since the advent of chromatography, the analysis of minute samples has become a routine matter. There are two widely used methods of collecting *small* samples of sieve tube sap. If a shallow cut (not reaching quite as far as the cambium) is made transversely into the young bark of a tree on a sunny day drops of liquid slowly emerge. They are almost certainly forced out from the sieve tubes, for if the cut goes a little deeper into the xylem they are drawn back owing to the negative transpirational pressure in the vessels. Zimmermann, Ziegler and others have collected this phloem exudate from quite a wide range of trees with some interesting results. In general, sucrose appeared as the major sugar of transport, sometimes reaching a concentration of 25% (about $\frac{2}{3}$ molar). Reducing sugars were invariably absent. However, with fair frequency oligosaccharides containing one or more galactose residues linked serially to sucrose were also present. In particular raffinose and stachyose were more or less equal in amount to sucrose in *Ulmus americana,* while stachyose exceeded it considerably in *Fraxinus americana.* No monosaccharides or sugar phosphates were found. Interestingly, the sugar alcohol mannitol was found in appreciable quantities in the phloem exudate of lilac *(Syringa)* and the manna-ash. Sorbitol is present in the apple.

Perhaps the most interesting method of sampling the sieve tube sap is to avail ourselves of the engineering skill of aphids. These insects commonly feed by inserting their minute stylets, often for quite considerable distances and by fairly devious routes, into individual sieve tubes. Once in this position the high turgor of the sieve tube ensures the insect without effort a continuous supply of food. If the aphid is very carefully anaesthetised it can be severed from its stylet and the latter left inserted into the stem. The exudate which with luck continues to flow from it can be collected in microcapillaries and analysed chromatographically. This method yields results in accordance with those of the previous methods; in *Salix, Heracleum* and *Picea* sucrose is the only sugar detected in quantity.

Two further lines of approach have emphasised the pre-eminent role of sucrose in translocation. Kursanov in Russia inserted a length of *Helianthus annuus* stem by double grafting into a plant of *H. tuberosus.* He also performed the reciprocal graft. As is well known *H. annuus* specialises in glucose and its polymers while *H. tuberosus* prefers fructose. On feeding a leaf with $^{14}CO_2$ and sampling the intercalated stem it was found that over 90% of the label was in sucrose whichever graft was being investigated. Evidently in both species transport is similarly in the form of sucrose. As another approach Trip and Gorham in Canada introduced radio-active glucose into the leaf of soya bean *(Glycine max)* through a lateral vein. It was first, they found, converted to sucrose and then transported from the leaf as such.

We can sum up therefore by saying that, no doubt to the satisfaction of this conference, sucrose is *par excellence* the form in which the products of assimilation are conveyed about the green plant. The reasons for this choice on the part of the plant are not however very obvious. In choosing a disaccharide rather than a monosaccharide the plant has reduced to one half the osmotic potential of its sieve tube sap; and as will appear later this may possibly be a very significant disadvantage. On the other hand sucrose is a considerably more stable compound than the hexoses, and ability to remain unmetabolised *en route* is an important advantage. It was mentioned earlier that invertase has not been found in the sieve tubes. As a disaccharide, sucrose has a higher free energy than the equivalent amount of hexose. It seems that the formation of the glucose polymer callose on the sieve plate is a very rapid and significant wound reaction, perhaps analogous to blood-clotting; and it may be that with its higher free energy sucrose converts more readily to callose than would glucose, and has been favoured on this account. Finally, in the absence of any agreed answer to the riddle of the mechanism of translocation we can speculate that perhaps sucrose has some special property that is significant for the mechanism, but which we are not yet in a position to recognise.

Before passing to a discussion of the pathway of movement a very significant fact must be mentioned. For some reason, perhaps only because of its osmotic preponderance, sucrose occupies a key position in the whole pattern of movement. If sugar is moving other things will move, inorganic ions, synthetic hormones, vitamins or viruses. But when there is no mobile carbohydrate the movement of these lesser components of the sap ceases. That is why selective weedkillers, for instance, must be applied in periods of rapid photosynthesis; it seems to be the sugar stream that bears them along.

THE PATHWAY OF MOVEMENT

The pathway along which it is now believed sugar migrates, the sieve tubes of the phloem, was in fact one of the earliest suggested; but it was a long time before sufficient evidence had been accumulated to make it the choice of an overwhelming number of plant physiologists. Other pathways mooted were the phloem parenchyma, especially the 'starch-sheath' consisting of large cells containing a heavy deposit of starch grains, and the young xylem vessels. One of the difficulties has for long been that with its numerous sieve plates the phloem tubes looked too resistive a path for the rapid movement of the assimilate stream. As we shall see, recent evidence seems to make the pathway very considerably more resistive still; it has therefore taken some very cogent evidence to convince physiologists that in spite of everything the sieve tubes are the conduits.

What lines of evidence point to this conclusion? In the first place there are the results of ringing experiments. It is well known that during the growing season the bark of many trees separates very easily from the wood along the cylinder of dividing cells called the cambium or rather, a little inside it. By this means the phloem tissue can be removed along a ring leaving the water-conducting xylem intact. When stems are ringed in this way certain well-established results follow of considerable use to horticulturalists: the roots slow down their growth, or if the ring is near the stem apex the tip of the shoot does so. Certain plants however (such as the Oleander) have a second set of phloem strands *inside* the wood of the young stem, and here the response is much less. Such facts as these obviously point to a food-conducting function for the bark tissues, though it can hardly be claimed that they indicate the sieve tubes specifically.

Mason and Maskell's classical work has already been alluded to. As a matter of fact they went even further than indicated. By dividing the bark tangentially into zones and analysing these separately they found that the diurnal fluctuation in sucrose concentration occurred principally in the zone with most sieve tubes. These therefore were implicated as the most likely conduits.

It is obvious of course that assimilate movement is not only *downwards* to the roots; expanding leaves in spring, the growing points of shoots, and developing fruits must also be supplied and these involve movement upwards. Could it be that when movement is upwards the pathway is other than when it is downwards, perhaps in the upwardly directed xylem stream? Ringing experiments partly answer this in the negative; another answer comes from the observation that steaming a portion of stem, depriving it of oxygen, or treating it with poisons, seriously interferes with assimilate movement or stops it altogether in whichever direction the presumed movement is occurring. This observation rules out the non-living xylem fairly conclusively.

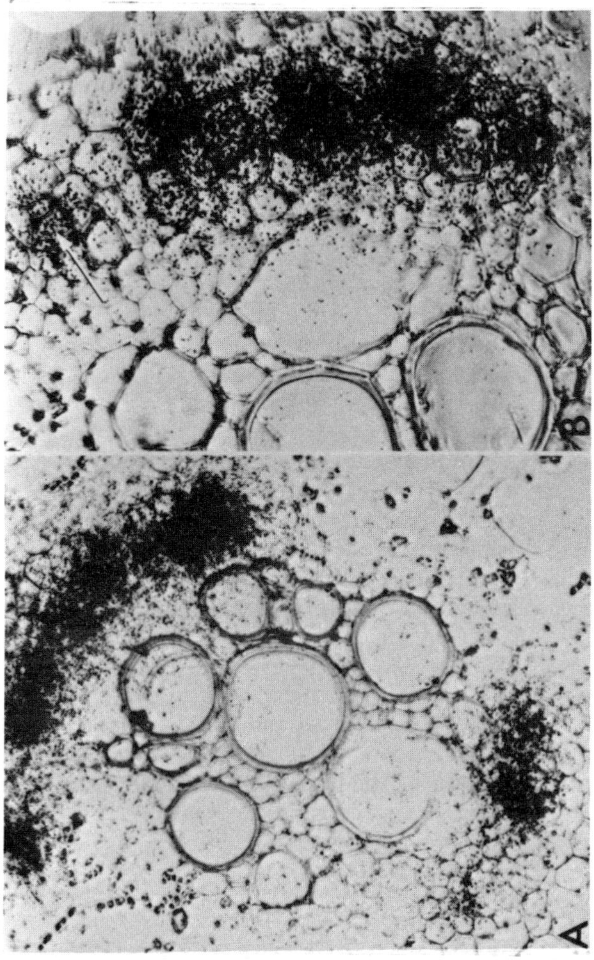

Figure 1. Autoradiographs of transverse sections of petiole of Concurbiba *following giving of* $^{14}CO_2$ *lo leaf. The labelled assimilate is clearly in the phloem (the bundles of* Cucurbiba *have two areas of phloem). The arrow in* B *points to an isolated sieve tube. The large cells are the xylem vessels. (By courtesy of P. Trip and P. R. Gorham,* Can. J. Bot. **45**, 1568, 1967*). Magnification ×430,* A, *×640,* B *(reduced six-tenths on reproduction)*

The fact that aphids insert their stylets almost always into sieve tubes has already been mentioned. Even more cogent however is evidence from tracers. Fluorescent dyes, like fluorescein, will enter leaves and pass into the phloem if placed in solution on the cuticle. They can then be observed microscopically in the living tissues of the petiole, and have actually been seen to be moving in the sieve tubes. Of course fluorescein is chemically very different from sugars, and the obvious next step was to use radio-active isotopes and see if sugar movement could be similarly localised. It had been found that the sugar being moved, as judged by the axial progress of the radio-active label, was sucrose; coupled with the known high concentration of sucrose in the sieve tubes (derived from aphid studies, often as we saw to the exclusion of all other sugars), this was evidence that these cells formed the conduit. However, actual visual localisation, as with fluorescein, was clearly desirable. This has now been achieved by a number of workers, first by giving $^{14}CO_2$ to the leaf and allowing it to photosynthesise, and later by applying tritiated sugars in solution directly to the leaf surface. Subsequent sectioning of the petiole or stem below the treated leaf, with freeze-drying and micro-autoradiography, allowed the labelled sugar to be localised firstly in the phloem bundles, and then actually in individual sieve tubes *(see Figure 1)*. It was the use of sugar labelled with tritium, whose weak β-emission gives very high resolution, that led to this final success of localisation in the sieve tubes themselves, as opposed to the companion cells or phloem parenchyma. Thus at last the intuitions of the morphologists have been confirmed by the experimental approach.

THE VELOCITY OF MOVEMENT

The very interesting question of the velocity of movement is bound up with that of the mechanism, as will appear later. It should be realised at the outset that there need not, in principle, be any single precise linear velocity attributable to a given case of translocation, any more than there is with diffusion, where the concentration front spreads itself out according to a probability function. What is quite precise and unequivocal however is a function called the specific mass transfer rate; that is, the amount of substance passing unit cross section of the phloem per unit time. If this is divided by the concentration of the substance (e.g. sugar) and by the fraction of the phloem tissue occupied by the actual conduits the result will be a linear velocity, and this may be looked on by convention as the linear translocation velocity at the point in question. In passing, the value of the specific mass transfer rate is about 3–5 g/cm^2/h in stolons and peduncles and considerably less in petioles.

One of the first and in many ways one of the best methods used this latter approach to ascertain the velocity. Dixon, who was one of the earliest to put forward an acceptable theory for the ascent of water to the tops of tall trees, recorded the increase in the dry weight of a potato tuber as it grew at the end of its stolon. By measuring the aggregate cross-section of the sieve tubes in the bundles of the stolon and by assuming a sugar concentration for the sieve tube sap, he was able, using his dry weight data, to calculate the average velocity of the phloem sap during the period of growth of the tuber. The result was about 50 cm per hour on the basis of a 10% solution. Similar calculations have been done for other species, such as yam (88 cm/h).

Another and very obvious method of measuring velocity depends on the use of radioactive tracers. In principle all one has to do is to apply a suitable tracer in solution to a leaf, or to give a leaf radioactive CO_2 and then time the passage of label down the axis. A more refined method depends on the following phenomenon. If one harvests the plant a short time after tracer has been given, cuts the axis up into equal segments and assays them for radioactivity, it is commonly found that the radioactivity falls off exponentially from the point of application. In other words, a plot of the logarithm of the radioactivity against distance gives a straight line. If the procedure is repeated for a duplicate plant but allowing the tracer to move for a longer time one obtains a straight line which lies rather further to the right *(see Figure 2)*. The horizontal intercept between these two straight lines can be regarded as the distance travelled by the assimilate stream in the time interval obtained by difference, and the velocity accordingly ascertained. It works out at about 30–100 cm/h. Strictly, a correction should be applied for leakage of tracer out of the conduits, but we can take it that broadly speaking the tracer method corroborates that obtained from mass transfer measurements and leads to the conclusion that translocation of sucrose occurs at a linear velocity of from 50–100 cm/h.

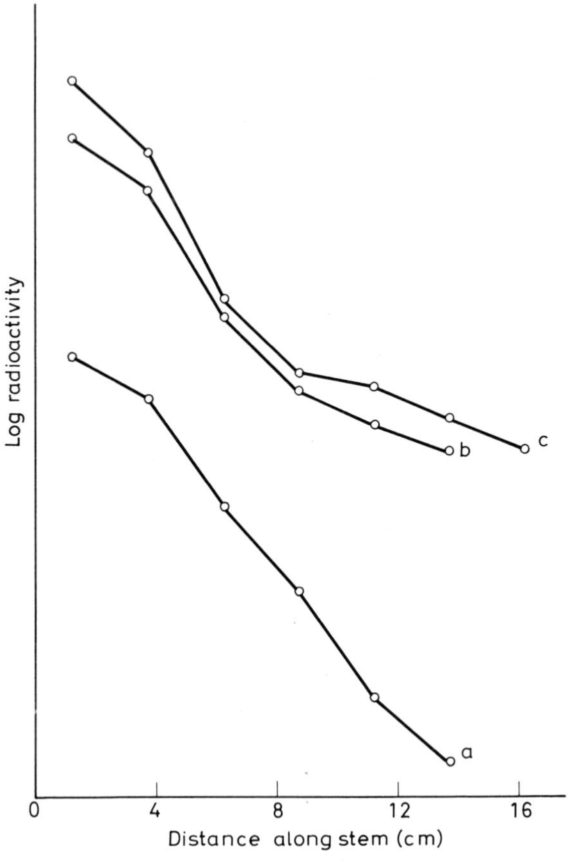

Figure 2. Movement of ^{14}C *tracer down the stem of* Phaseolus. *The stem below the node of the treated leaf was cut into* 2·5 cm *sections and assayed for radioactivity after the leaf had been given* $^{14}CO_2$. *The delays before harvesting were* (a) 15 min, (b) 20 min, (c) 30 min. *(Redrawn from O. Biddulph and R. Cory by courtesy,* Pl. Physiol. 32, 612, 1957*)*

Before leaving the subject of velocity, two further points may be noted. Owing to the alternation of day and night there is a 24 h period in the sugar concentration in leaves, and therefore of the intensity of translocation. This results in waves of sucrose concentration passing down the trunks of trees. If the wavelength of these is measured this will represent the distance moved in 24 h and will yield a velocity of movement. This has been done by Huber, and the result is rather high. It may in fact reflect also changes in water tension in the xylem, and Zimmermann has recently repeated the observations using the ratios between sucrose, stachyose and raffinose in the trunk of *Fraxinus americana,* instead of sugar concentrations. His results for the velocity are much lower, and in fact fall well within the range previously quoted.

The other point concerns an appreciation of the velocity in relation to the size of the conduits. Some idea of this relationship can be gained by noting that under a 4 mm objective with a ×10 eyepiece the field of view would be traversed by the assimilate stream in about a second and a half. Otherwise expressed, the sap stream would move along the length of a sieve tube element in from 2–4 seconds. On the microscopic scale, therefore, and in relation to the size of the conduits the speed of movement appears quite high.

THE STRUCTURE OF THE PHLOEM

We can summarise what has gone before by saying that the sieve tube system of the plant constitutes a network of conduits along which moves a solution commonly containing 10–25% sucrose with much smaller amounts of nitrogenous compounds and ions. The velocity of flow may be anything between 30 and 120 cm/h, and movement may be upwards or downwards according to the relative positions of sources of supply (leaves or germinating storage organs) and centres of demand (root and shoot tips, developing fruits, etc.). In general, flow is between sources and sinks closest together (as in a plumbing system). This means that in a given stem flow may be in one direction at one point and in the opposite direction further along. Even at the same cross-section it may be differently directed in adjacent bundles, depending on the detailed vascular anatomy of the axis.

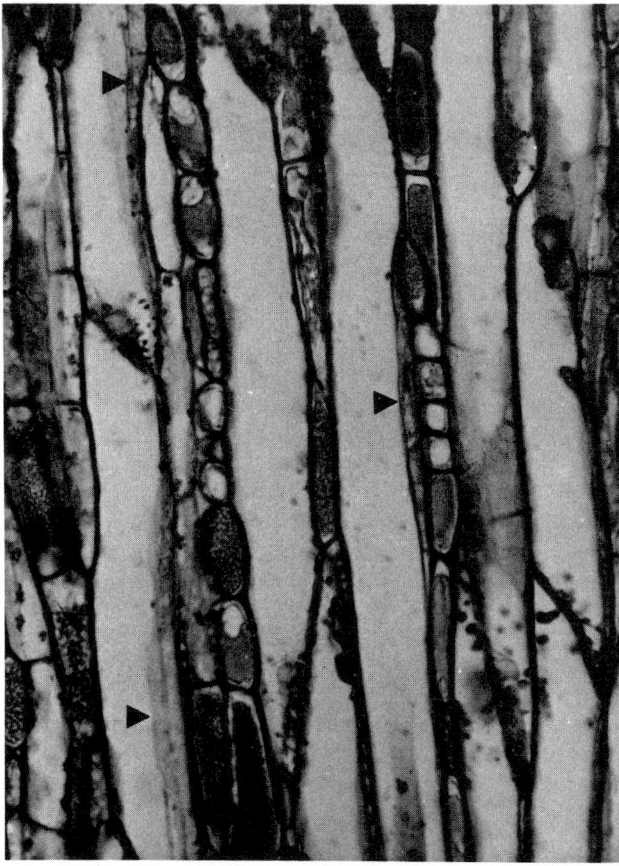

Figure 3. Tangential longitudinal section of the phloem of Salix caprea. *The large cells are the sieve tube elements, the oblique end walls being the sieve plates. The narrow tapering companion cells can be seen alongside the sieve tubes (see arrow heads). (By courtesy of U. Mishra). Magnification ×900 (reduced six-tenths on reproduction)*

Before we discuss the question of mechanism it will be helpful to recall the main features of the phloem structure as described by the classical anatomists and to make some remarks about more recent revelations of the electron microscopists. Besides a common content of parenchyma cells and bundles of fibres the phloem is made up of sieve tube elements and their associated companion cells. Typically, sieve tube elements are elongated cells (from 0·2–1 mm long and one-twentieth as wide) placed end to end and with their cross walls partially broken down to form the well-known sieve plates *(see Figure 3)*. These sieve plates may be fairly accurately transverse, or very oblique; in either case they are perforated with definite holes which may be anything from $\frac{1}{4}$–4μ in diameter. The cells unlike the xylem vessels retain their living character, though their cytoplasm is very sparse. They have however lost their nucleus and with it most of their organelles including the tonoplast. A few mitochondria remain, and rather more somewhat unusual plastids. The general appearance of the protoplast however conveys the impression of metabolic lassitude. To make up for this, sieve tube elements are very closely associated with cells having dense cytoplasm, a large nucleus, and abundant ribosomes, the so-called companion cells. These arise from the same mother cell, and may be one or several in number. They are connected by specialised plasmodesmata to the sieve elements, and seem to have some important auxiliary function.

In the development of the sieve elements there always appears, by some process or other, a dark-staining substance which has traditionally been referred to as 'slime'. As the cell matures and its cytoplasm becomes more sparse the slime disperses, and it is only in fairly recent times that the electron microscope has drawn our attention to the fact that it has a well-organised fine structure and appears to fulfil some important function in translocation. To begin with, the mature sieve tube element is lined by a well-marked plasmalemma that responds to plasmolysing solutions, and so can be regarded as fairly impermeable to solutes like sucrose. This plasmalemma lines the pores in the sieve plate; that is, it does not lie over the sieve plate and separate one element from the next *(see Figure 4)*. The sieve plate pores would thus constitute open connections were it not for the very puzzling circumstance agreed on by most electron microscopists—they are more or less blocked by slime substance. This substance is shown to consist of extremely fine fibrils, probably protein in nature and about 10 nm in diameter. In good preparations they are most usually seen with a fine banded, periodic structure, something like prepared fibrils of collagen, the period like the diameter, being also about 10 nm *(see Figure 5)*. These fibrils lie axially through the

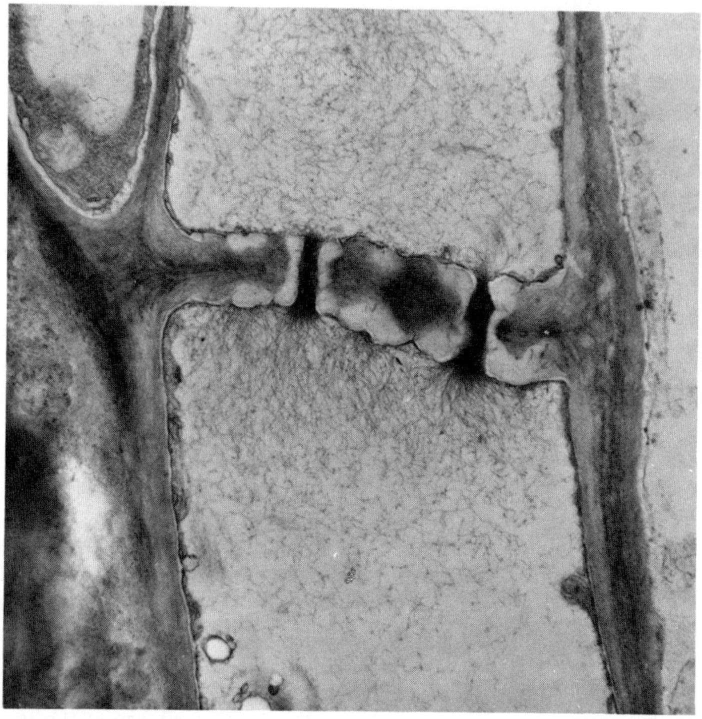

Figure 4. Electron micrograph showing longitudinal section of a sieve plate from Helianthus annuus. *Note the slime fibrils in the lumens and filling the sieve plate pores. The white areas on the plate are callose, some of which is formed during fixation. The small cell to the side is a companion cell. (By courtesy of A. W. Siddiqui). Magnification* × *25 000 (reduced one half on reproduction)*

sieve plate pores, commonly fraying out into the lumens of the sieve elements above and below the plate and frequently expanding into a loose fibrillar meshwork in the lumen. Sieve tubes possess a mechanism for very rapidly depositing a glucosan called callose on the sieve plates whenever their conducting function is interfered with. This reaction proceeds even in the presence of the usual electron microscopic fixatives like glutaraldehyde; and consequently most electron micrographs of the sieve plate show a thicker than normal ring of collose around the periphery of the pores. This squeezes the slime plugs to a smaller diameter, and almost certainly gives the impression that they are denser than they are in the functioning state. Thus we may tentatively assume that the functioning sieve tubes possess a very thin lining layer of cytoplasm with few organelles, and that their large lumens communicate freely through the sieve plate pores

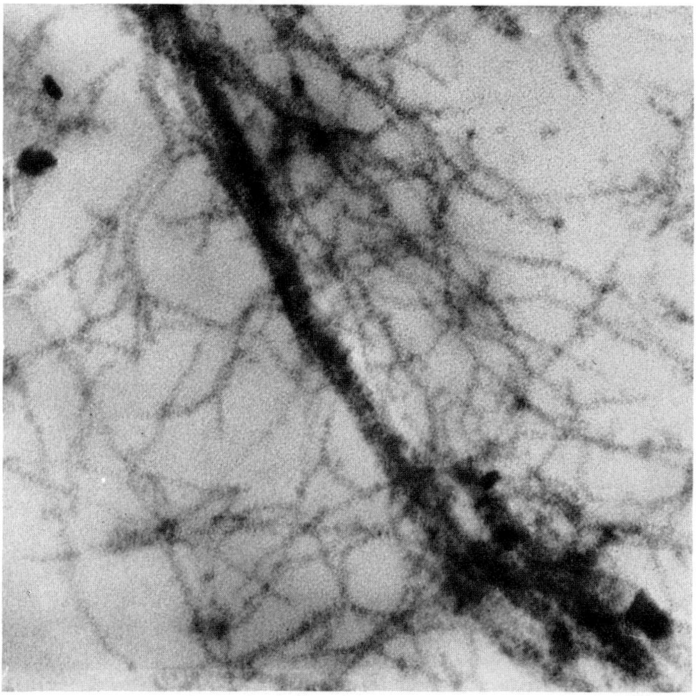

Figure 5. Slime fibrils from a sieve tube of Salix. *Note the banded appearance. (By courtesy of U. Mishra). Magnification ×150 000 (reduced one half on reproduction)*

except for loose plugs of fine parallel protein fibrils which occupy the latter positions.

The presence of the fibrils in the pores constitutes so great a difficulty in understanding how translocation occurs that strenuous attempts have been made to establish that their appearance there is an artefact of tissue preparation. Sieve tubes with their high sugar content have a large internal turgor pressure. When they are cut into therefore there is a sudden release of pressure; and this, it has been suggested, displaces the fibrillar material into the pores. Or again, it has been suggested that just as callose is laid down quickly on interference with the phloem so the fibrillar plugs might be enzymatically formed *in situ*. Both of these suggestions seem to be extremely unlikely, but they do emphasise the theoretical problem. This is not chiefly that the fibrillar material seriously occludes the pore on a cross-sectional area basis; but rather that by narrowing down the flow channels it adds very greatly to the frictional resistance to flow. Poiseuille's equation, which gives the flow rate under pressure through a circular pipe, is relevant in this connection. It can be written in the form:

$$\text{mean linear velocity} = \frac{r^2}{8\eta} \times \text{pressure gradient}$$

where η is the viscosity of the moving fluid. The equation indicates that as the radius r decreases the pressure gradient needed to maintain a given velocity of flow goes up sharply (in fact inversely as the square of the radius). The channels between the fibrils may be far from circular tubes of course, but the principle remains the same; subdivision of the flow path into narrower channels seriously impedes the movement. It is this which faces the student of sieve tube mechanism with one of his greatest problems.

THE MECHANISM OF TRANSLOCATION

We are now in a position to consider the various mechanisms which have been suggested to account for the rapid movement of the assimilate stream through the sieve tubes. Four possibilities will be discussed.

One of the earliest invoked the well-known phenomenon of protoplasmic streaming. It needs to be remembered that the measured rate of sugar transport is many thousands of times greater than could be achieved by simple diffusion down a concentration gradient. But if cyclosis, such as can be observed in many plant cells, operated in the sieve tubes it could possibly make it permissible for diffusion to be invoked only over the sieve plate with cyclosis taking over down the lumen. The trouble with this suggestion was that until

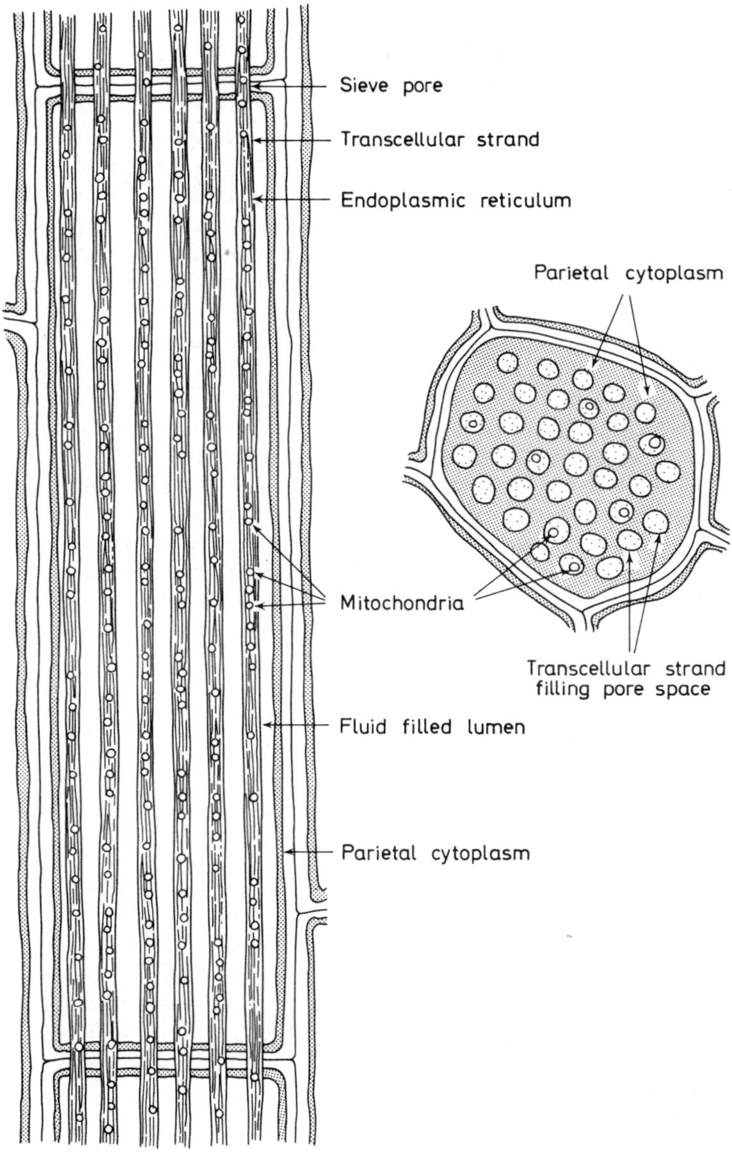

Figure 6. The transcellular strand theory of R. Thaine by courtesy (J. exp. Bot. 15, 474, 1964).

recently protoplasmic streaming had never been observed in mature sieve tubes, though it had been in immature ones. Also the velocity of streaming in vascular plants is considerably less than that of translocation (say one-twentieth), and if diffusion has to be credited with movement through only the sieve plate the overall rate which can be achieved is nevertheless quite inadequate. Within the last two years, however, using newer optical techniques such as phase contrast, Thaine and others have observed and photographed movement in functioning sieve tubes which they have attributed to an elaborate system of cytoplasmic tubules running longitudinally between and through the sieve plates *(see Figure 6).* They call these 'transcellular strands', and suppose that a rapid transport of sugar occurs within them motivated by some sort of peristaltic or other action. This theory, which is quite popular, runs into the difficulties that nothing like these membrane-bound strands has ever been seen in the electron microscope; and that movement for the full length of the sieve tubes in such very narrow tubes (less than the diameter of the sieve plate pores) would involve a very great increase in energy expenditure. Further, the optical microscope images which Thaine and others have seen and which

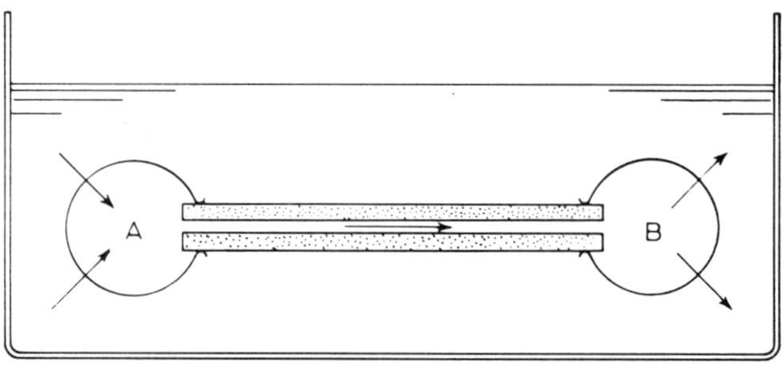

Figure 7. Diagram to illustrate the pressure-flow theory of Münch. The dish contains water. The bladders at the ends are semi-permeable, A *containing strong sugar solution and* B *weak. The inclined arrows indicate movement of water. the horizontal arrow movement of solution.*

constitute the principal evidence for the theory have been interpreted in other ways. Until some evidence of an electron micrographical character has been produced for these strands it is difficult to believe that these alternative interpretations are not the correct ones.

Passing over several very vague suggestions that somehow movement of sugar is by some sort of metabolically activated diffusion, the next mechanism to be considered is the classical Münch or pressure-flow hypothesis. This suggests that movement down the sieve tubes is driven by a simple gradient of hydrostatic pressure continuous from source to sink. The mechanism is illustrated by *Figure 7.* So long as a difference of osmotic potential exists between the ends A and B, flow will occur down the connecting tube conveying solute with it. In terms of the plant, A represents the sieve tube elements adjacent to the photosynthesising cells of the leaf or the storage cells of the shrinking tuber, while B represents the sieve tubes ending in the growing points. Sugar is being loaded into the sieve tubes at the supply end and unloaded from the sieve tubes at the demand end, in both cases probably by some method involving phosphorylation. The difference in concentration at the two ends sets up a difference in turgor pressure, and this constitutes the driving force for flow.

Münch's hypothesis is probably the one most favoured at the present time. It invokes no physical principles of doubtful validity and accounts for a great deal of the observed behaviour of the phloem. However, of late it has run into some damaging criticisms, chief among which is the difficulty occasioned by the plugging of the sieve plate pores by slime fibrils which we noted earlier. It is not difficult to calculate (though the calculation involves extrapolating the laws of macroscopic flow into the microscopic region) that the pressure gradient needed to propel solution through the columns runs into very high values, perhaps hundreds of atmospheres per metre. The resistance is located chiefly at the sieve plates. Needless to say unless the slime-plugging is an artefact, or some quite new physical principles are involved in microscopic flow, this difficulty is a fatal one. The Münch hypothesis also fails to do justice to the effect of heat, poisons or oxygen deprivation, and to the considerable evidence for simultaneous flow in opposite directions, at least in the same vascular bundle. Moving the assimilate stream means of course doing work in the thermodynamic sense. Somewhere therefore a source of free energy, probably metabolic, must be involved. In the Münch hypothesis this happens exclusively at the ends, where sugar is loaded and unloaded against concentration gradients between sieve tubes and parenchyma cells. There is no need for metabolic energy supply *en route,* except for the purpose of 'care and maintenance' of the conduits. Measurements of the respiration of phloem however have shown that it is fairly high, considerably in excess of what such care and maintenance would seem to require; and this constitutes one of a number of lesser difficulties for the theory.

The objections to the Münch hypothesis have led to the development of theories involving bulk flow (i.e. mass flow) in the sieve tubes actuated not passively by a pressure-gradient but actively by some sort of pumping action localised at the sieve plates. The advantages of such theories over the Münch hypothesis are that they utilise energy released *en route*; they supply a reason for the frequent multiplication of sieve plates in the development of sieve tube elements from the cambial initials; they fairly readily accommodate simultaneous two-directional movement in the same bundle; they allow for the effects of poisons or oxygen deprivation; and they easily surmount the problem of plugging of the sieve plate pores. In addition they can offer explanations for other unusual features of the phloem such as its high content of potassium, not to mention the ubiquitous presence of the fibrillar slime.

It is indeed on this last factor that the active mass-flow theories principally depend. For what can be the possible significance for translocation of this strange constituent of the sieve tubes? It has been suggested that it is a storage protein; after all, carbohydrate is conspicuously present in the sieve tubes in the form of the peculiar starch granules occurring in their plastids, as well as in the form of thick cell walls and callose. But in answer to this it hardly seems necessary for a merely storage protein to be dispersed in such a flow-obstructing form as the slime fibrils. In certain families, such as the Papilionaceae, the immature sieve tubes contain dense protein crystals which disperse as the elements mature into a hairy mass of fine fibrils forming the slime. Were protein storage the function, one would hardly expect this to be the course of events. Then it has been suggested that the fibrillar matter helps to ensure rapid sealing of the sieve tubes in the event of damage. This may well be so, but if this is all, it seems a big price to pay for a function which appears adequately catered for by the callose mechanism, which utilises the cheaper and more abundant carbohydrate. In any case, both of these functions (storage and sealing) may be valuable without constituting the *raison d'être* of the slime substance.

Two other suggestions have been put forward. One is that the fibrillar material is in fact a contractile protein, which in the presence of the apparently high ATP content of the sieve tube sap contracts rhythmically in such a way as to propel the sap along. Certain appearances under the optical microscope have been appealed to in this connection. It is however very difficult to see how fibrils, without apparent anchorage (compare cilia) and with little spatial organisation, can act in this way. Their orientation in the mature element seems fairly random except in the sieve plate pores, and even here there is little to suggest organisational relationship. How do they manage, on different occasions, to propel sap in both directions as the known lack of polarity of the sieve tubes implies that they must do? How do they receive the signals that make them relax or contract (in harmony) when surrounded by a rapidly moving milieu? The analogy with cellular organisation in the animal world seems too remote to be pressed very far.

A second suggestion is that the fibrillar material in the pores constitutes an electro-osmotic framework. Electro-osmosis is a common, well-understood and powerful phenomenon. It occurs whenever a porous material immersed in a dilute solution of an electrolyte is subjected to an electrical potential gradient *(see Figure 8)*. Glass wool, fine sand or cellulose membranes show it well. The mechanism works as follows. The porous material if suitable possesses fixed electrical charges on its walls, due either to ionisation (as in the case of protein) or to adsorption (as with fine quartz sand). The minute channels, which are much larger than the holes in a semi-permeable membrane but are still very small, are influenced by the electrical fields of these fixed charges. As a result the water columns within them contain more of the mobile ions of one sign (the opposite one) than of the other. When an electrical gradient is imposed across the porous material all the mobile ions it contains move; but there being more of one sign than of the other, the minute water columns are dragged unequally in the two directions, and a net movement results. The requirements for electro-osmosis are therefore these: a membrane with suitably fine porosity composed of a substance of medium to high surface charge; an ionic mileu of strength 0·1–0·001 normality; and a sustained difference of potential between the two sides. It is possible to view the sieve plate in this light. The fibrillar material is probably protein in nature and stains deeply with electron stains such as the lead ion. It would therefore seem to carry charged groups, probably negative, on its surface. The structure of the plugs in the sieve plate pores appears to be adequately fine. The sieve tube sap has a content of about 0·03 N in potassium, this being the only abundant cation. A

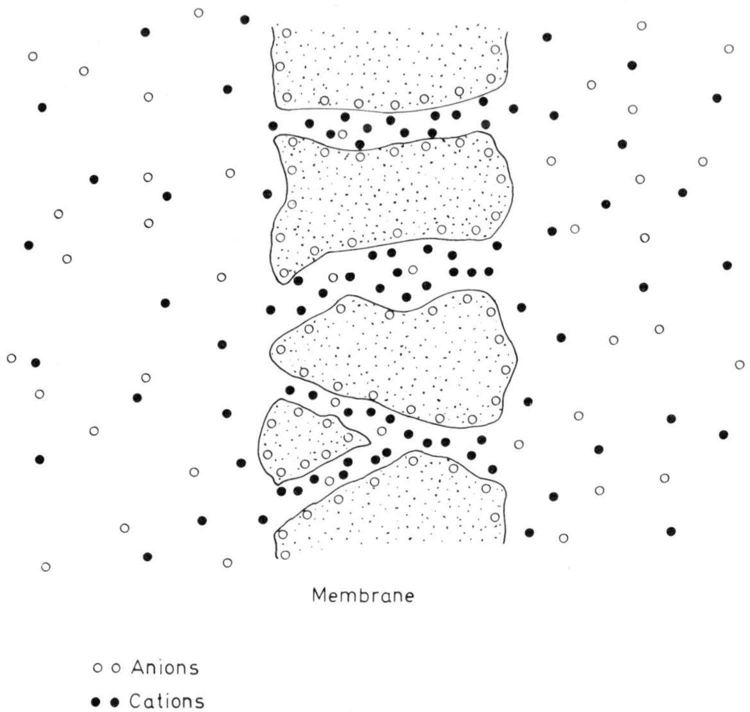

Figure 8. The mechanism of electro-osmosis through a porous membrane. When a potential difference is imposed across the membrane the solution in the pores undergoes transport as a consequence of its unequal content of anions and cations.

potential difference across the sieve plates, facing the same way all down the column, was originally postulated on the supposition of an active uptake mechanism for potassium associated with the companion cells. Such a saw-tooth profile in the potential, with steps of about 20 mV per sieve plate, has now actually been observed in the phloem of the grape vine. The suggestion therefore is that potassium is transported by the cell membrane into the sieve tube above the sieve plate, and that it then moves down the electrical gradient across the plate dragging sap with it. Underneath the plate it is leaked out by the cell membrane and by some route moves back to its initial position. Thus the plate acts as an electro-osmotic pump, the energy to keep it running coming from the metabolism associated with the active uptake of the ion. In this scheme therefore the slime fibrils fulfil an essential role in providing a finely-grained framework of fixed negative charges, without which electro-osmotic forces could not be developed. The theory faces considerable difficulties of its own; in particular it seems to call for a very high potassium current and for a pathway for return flow for which there is yet little experimental evidence. But it does offer a reasonable explanation for the presence of the slime substance and for the means by which the resistance it entails is overcome.

CONCLUSION

This account may be suitably closed by drawing together some of its main arguments. The leading role in conveying organic assimilate about the plant has been entrusted in the main to sucrose, possibly on account of its chemical stability as compared with hexoses. Sugars hold a dominant position in the whole transport business of the plant; other substances, present in much smaller quantity, move rapidly only when sugar is on the move. The channels of transport are the sieve tubes of the phloem, and the linear velocity of movement can be quoted as 30–120 cm/h. Recent electron microscopical evidence has shown that the channels are probably very resistive, and at this velocity require more effort to promote movement than a simple pressure gradient can supply. At the moment there is no generally accepted theory as to the motive forces involved, though several hypotheses are being canvassed. Despite the problems for thought, however, the plant seems to manage the process very successfully.

BIBLIOGRAPHY

This account has quoted very few authorities. For those wishing to pursue the matter further, and to consult original papers, the following reviews may be recommended:

ESAU, K., CURRIER, H. B. and CHEADLE, V. I., 'Physiology of Phloem', *A. Rev. Pl. Physiol.*, **8**, 349–374 (1957)

STEWARD, F. C., *Plant Physiology—a Treatise,* Vol. II, articles by C. A. Swanson and O. Biddulph, Academic Press, London (1959)
WEATHERLEY, P. E. and JOHNSON, R. P. C., 'The form and function of the sieve tube: a problem in reconciliation', *Int. Rev. Cyt.,* **24,** 149–192 (1968)
ZIMMERMANN, M. H., 'Absorption and Translocation: Transport in the Phloem', *A. Rev. Pl. Physiol.,* **11,** 167–190 (1960)

Chapter Thirteen

PLANT GROWTH REGULATORS AND THE ACCUMULATION OF SUCROSE IN SUGAR CANE

D. W. FEWKES
Head of Biology-Agriculture Department, Tate & Lyle Ltd, Research Centre, Keston, Kent

INTRODUCTION

Under normal conditions of cultivation sugar-cane plants synthesise more carbohydrate than is required for vegetative growth. Modern sugar-cane varieties are species and hybrids of the genera *Saccharum* and *Erianthus* which store this excess carbohydrate in their stalks mainly as sucrose. The exploitation of this characteristic by plant breeders has resulted in the selection of high sucrose yielding varieties.[1]

Apart from such considerations as erectness (to facilitate mechanical harvesting) and resistance to diseases or insect pests, new commercial sugar-cane varieties are chosen for their ability to produce high yields of sucrose. The two main yield components are weight of cane per unit area (number of stalks and mean weight per stalk) and extractable sucrose per cent cane. Thus equivalent yields per acre are obtained from 40 tons of cane containing 10% sucrose and 25 tons of cane with a 16% sucrose content. From economic considerations the latter combination is preferable because the harvesting, transport and processing costs are lower whilst cultivation costs are much the same.

FACTORS AFFECTING SUCROSE ACCUMULATION

The total amount of carbohydrate available for growth and storage depends upon the photosynthetic capacity of the sugar-cane plant and upon environmental factors which influence the rate of photosynthesis. The partition of photosynthetic assimilates between vegetative growth and storage is also affected by environmental factors. When temperature, moisture and nutrient levels are optimum for vegetative growth a relatively small proportion of new assimilate enters into storage. Under conditions less favourable for vegetative growth proportionately more sucrose is stored. However, environmental conditions which favour a high rate of photosynthesis may also favour rapid vegetative growth. There is obviously an ideal situation in which the ratio of sucrose to non-sucrose materials and the weight of cane per unit area both reach high levels at harvest.

The bulk of a sugar-cane crop at harvest consists of mature stem tissue. The main organic constituents of this storage tissue are carbohydrates: sugars and cell wall materials. It is evident that growth and storage are inversely related because they are in competition for the carbohydrate produced during photosynthesis. Therefore those factors which affect the growth of sugar-cane also affect sucrose accumulation.

The factors which control plant growth all originate as interactions between the genetic constitution of the plant and its environment. Genetic factors are paramount in that they determine the nature and extent of a plant's response to the environment and thus set limits within which environmental influences may find expression. This interplay between genetic and environmental factors is mediated by hormonal activity.

Some environmental factors exert a more-or-less direct influence upon plant growth, as for example the general effect of temperature upon metabolic activity. Others, such as soil moisture or light, impinge upon one part of the plant but also affect growth processes in another part. The transmission of such environmental 'messages' within the plant is effected by plant growth hormones whose activity and synthesis is sensitive to environmental influences. The hormones themselves act mainly at the genetic level to stimulate or inhibit enzyme synthesis. In this way they influence the distribution and the fate of organic substances synthesised by the plant, and thus co-ordinate plant growth.

THE REGULATION OF PLANT GROWTH

Four groups of naturally occurring plant growth hormones are recognised: auxins, gibberellins, cytokinins and growth inhibitors. These substances occur in the plant as free acids or bases and also as conjugates in which the acid or base is associated with a sugar moiety such as glucose or ribose. The free acids and bases are moderately water soluble but the linkage with a sugar enhances this. It is generally accepted that the free acids and bases are the active forms and that the conjugates are probably inactive but mobile, easily transportable forms of the plant growth hormones.

The gas ethylene, which also has hormonal activity and occurs naturally in plants, has long been regarded as a plant growth hormone.[2] Ethylene has been shown to promote the ripening of fruits,[3, 4] accelerate the senescence and abscission of leaves, fruits and flowers[5] and inhibit stem elongation.[6] Its production in plants is induced at certain critical concentrations of auxin which vary in different tissues, and this is thought to explain the growth inhibitory effect with supra-optimal levels of auxin.[6]

The mode of action of plant growth hormones is not understood although it seems likely that their main site of action is close to the gene.[7] The inherited characteristics of each species are coded as genes in its chromosomes, long, paired strands of deoxyribonucleic acid (DNA) present in the nuclei of its cells. Throughout the life of each cell, genetic information from the DNA strands is transcribed into ribonucleic acid (RNA) 'messengers', which pass out of the nucleus to the ribosomes where amino acids are linked to form enzymes according to the sequences coded in the RNA messengers. The administration of appropriate hormones to plant or animal tissues has been shown to enhance RNA synthesis, leading to increased protein synthesis and enzymatic activity.[8–12] Co-administration of substances which specifically inhibit DNA activity (e.g. puromycin, actinomycin D) prevents this response.[13–15] Thus, in some instances at least, plant growth hormones are involved in the earliest stages of enzyme synthesis.

Although each plant cell contains the complete genetic information of its species, only part of the genetic code is required for each stage of its development and different parts are required for cells destined to perform different functions. Bonner[16] considers that plant and animal hormones are 'effector' substances which activate genes previously repressed, thus causing an increased synthesis of messenger RNA (mRNA) molecules and, consequently, the synthesis of new kinds or increased amounts of enzyme molecules. While this may be true in some instances, there is also evidence that plant growth hormones influence a later stage in enzyme synthesis. Gayler and Glasziou[17] found that in sugar-cane immature internode tissue, the short-term response of enzyme synthesis to the addition of plant growth hormones was not due to increased mRNA synthesis, but rather to an increased rate of enzyme production from already formed mRNA. They postulate that auxin and gibberellin stabilise the mRNA for invertase in this tissue.

THE GROWTH HORMONES OF SUGAR-CANE

Sugar-cane contains representatives of the four main groups of plant growth hormones. Auxins have been isolated from stem apical tissue,[18] from the roots[19] and from the true seed[20] of sugar-cane. Gibberellins are found in extracts and diffusates of immature stem and leaf apical tissue,[21–25] in bleeding sap from cut stumps of sugar-cane stems,[26] in the roots[27] and in dormant and developing lateral buds.[24, 26, 28] Cytokinins also occur in bleeding sap and in dormant lateral buds.[29] Recently the growth inhibitor, abscisic acid has been isolated from immature stem and leaf apical tissue together with several other unidentified growth inhibitors.[30] Auxin (probably indole-3-acetic acid) and abscisic acid have been found so far only as free acids, but gibberellins and cytokinins have been found to occur in both the free and the conjugated, readily water soluble forms.

The simultaneous occurrence of all four types of plant growth hormone in stem apical tissue indicates that each one has a function to perform in stem growth, and suggests that the regulatory mechanism may involve changes in the hormonal balance. It is characteristic of the hormonal regulation of plant growth that more than one type of growth hormone is involved in any one growth process and that their functions often overlap.[31] Thus, auxins and gibberellins are both involved in the promotion of cell elongation, auxins and cytokinins in the stimulation of cell division. However, in different situations the actions of the same two hormones may be similar or opposed. For example, the presence of cytokinin and auxin enhances the growth of callus in tissue culture,[32] but these hormones oppose one another in controlling the germination of lateral buds.[33]

In immature stem tissue the presence of auxin appears to be essential for the cell elongation and cell division promoting activities of gibberellins and cytokinins. Abscisic acid (ABA), on the other hand, inhibits these processes. In simple terms, one might visualise auxins as having a 'juvenilising' effect upon plant cells, predisposing them to divide and grow in response to cytokinins and gibberellins, whilst ABA inhibits the response to these hormones and tends to 'age' the cells. For properly co-ordinated growth a balance between the effects of auxin and ABA is necessary; some cells must divide and some must grow, whilst others mature into their final form. This balance could be held by a differential rate of synthesis between auxin and ABA, and might also be influenced by gibberellins and cytokinins which are known to counteract the inhibitory effects of ABA in various plant systems.[34–38]

Most's discovery that the bleeding sap of sugar-cane is rich in water-soluble gibberellins and cytokinins, together with his isolation of ABA from immature stem and leaf apical tissue, led him to hypothesise that stem elongation in sugarcane is regulated by environmentally induced variations in the amounts of gibberellin and cytokinin translocated from the roots to the stem apex.[39] According to this theory, when the supply of gibberellin and cytokinin to the stem apex is plentiful the growth inhibitory effect of ABA is largely counteracted and growth is rapid; when this supply is restricted the growth rate is reduced. There is evidence from other plants that gibberellins and cytokinins are synthesised in the roots[40, 41] and translocated in the xylem.[42, 43] Also, Vaadia and Itai[44] report that the application of a mild water stress to the roots of sunflower causes a rapid reduction of cytokinin synthesis in the roots, long before photosynthesis is affected or wilting becomes apparent.

In wheat, the application of a more pronounced moisture stress, evidenced by wilting, causes an increased production of ABA in detached leaves.[45] Most[46] has also found that there is an increase of growth inhibitors in the leaves of moisture stressed sugar-cane plants. This may have some bearing upon Hudson's observation that at any particular moisture deficit over a fairly wide range, growth rates of sugar-cane were nearly twice as high in dull weather, compared with growth when the weather was sunny and dry.[47] Thus at low soil moisture deficits, which may not inhibit the synthesis of cytokinins and gibberellins in the roots, moisture stress in the aerial parts of the plant induced by hot, dry weather may inhibit growth through the enhancement of ABA synthesis in the leaves.

The role of auxin as probably the main co-ordinator of plant growth is generally accepted. This hormone is synthesised in the young leaves or the actively growing stem apical tissue and diffuses downwards throughout the plant, exerting an inhibitory effect upon the growth of lateral buds. Decapitation of the plant stem removes this source of auxin and releases the lateral buds from apical dominance, allowing them to develop into side shoots. Apical dominance appears to be due, in part at least, to auxin-induced production of ethylene,[48] and can be counteracted by injecting cytokinin.[33]

Roots are also sensitive to auxin, and at levels which normally promote stem elongation auxin inhibits root growth.[49] Cutler and Vlitos[19] found that the amount of extractable auxin in sugar-cane roots was negatively correlated with the growth rate of the aerial parts of the plant over a 12-day period prior to extraction. Perhaps when stem growth is slow increased amounts of auxin may be free to diffuse down the stems into the roots and exert a retarding effect upon their growth.

Supra-optimal levels of auxin tend to inhibit gibberellin induced stem elongation,[50] and Most[25] has found that immature stem apical and leaf tissue from slow growing sugar-cane appears to contain a higher level of auxin than similar tissue from rapidly elongating cane. In slow growing cane the immature internodes become lignified more rapidly and end up shorter than those of fast growing cane. The auxin indole acetic acid (IAA) is an anti-oxidant and a powerful inhibitor of lignin synthesis, but it also promotes the synthesis of peroxidases.[51] High levels of IAA in plant tissues induce the release of ethylene and this gas also promotes the synthesis of peroxidases.[52] Siegel[51] suggests that although IAA may temporarily suppress the oxidation of metabolites and enzymes essential for growth, it also tends to increase the peroxidising power of the tissues and hence its own destruction. When this occurs together with relatively low levels of gibberellin and cytokinin the tissues may tend to age prematurely under the influence of abscisic acid, which is known to promote the synthesis of enzymes involved in the lignification process.[53]

GROWTH HORMONES AND INVERTASE ACTIVITY IN SUGAR-CANE

In sugar-cane, carbohydrate is translocated and stored mainly as sucrose.[54] In stem tissue sucrose is inverted prior to its utilisation in metabolic processes, and it is significant that auxin (α-naphthalene acetic acid), gibberellic acid and abscisic acid all stimulate invertase synthesis in this tissue.[17] Variations in the levels of several invertases have been correlated with the movement and utilisation of sucrose.

During their accumulation in sugar-cane stem tissue sugars move through three distinct cell compartments, known as the outer space, the metabolic compartment and the storage compartment.[55] Sucrose is inverted during its passage from the conducting tissue into the storage compartment (cell vacuole), where it reappears as sucrose. This inversion takes place in the outer space (cell wall) and is mediated by an acid invertase. Resynthesis to sucrose takes place during the passage of sugars through the cytoplasm (metabolic compartment) to the vacuole.

In immature stem storage tissue a second soluble acid invertase occurs in the cell vacuole, and the synthesis of this enzyme is stimulated by auxin.[56] The storage compartment acid invertase appears to be concerned with the hydrolysis of sucrose prior to its return from storage to be used in growth. Hatch and Glasziou[57] found a correlation between total acid invertase activity and the rate of elongation of internodes when sugar-cane plants were subjected to different temperature or moisture stress regimes. A similar correlation was found for sugar-cane plants treated with different rates of gibberellic acid.[58]

In mature storage tissue of *Saccharum officinarum* L. varieties and hybrids thereof which have a high capacity for sucrose accumulation, the soluble acid invertases are replaced by a soluble neutral invertase, located in the cytoplasm, and an acid invertase which is firmly bound to the cell wall.[59] In contrast, soluble acid invertase activity was found in mature stem tissue of the wild canes *Saccharum spontaneum* L. and *Erianthus* sp., which accumulate little sucrose.[57]

Hatch and Glasziou[57] found that the sucrose content of mature internodes is closely related to neutral invertase activity. Both tend to increase when sucrose is moving into storage in response to low temperatures or moisture stress, and to decrease when the stem is elongating rapidly. The neutral invertase appears to be involved in the passage of sucrose from vascular tissues into the storage tissue of mature internodes.

RIPENING OF SUGAR-CANE

Accumulation of sucrose by the sugar-cane plant occurs throughout the major part of its growth cycle, from the time that storage tissue is first elaborated until harvest. Enhanced accumulation, known as 'ripening', occurs during the cooler and drier times of the year when vegetative growth is relatively slow. Normally, the crop is harvested during this period.

Where sugar-cane is grown under irrigation, ripening may be induced by reducing the supply of water to the crop at an appropriate time although the degree of ripening also depends upon other environmental factors, particularly solar radiation and day and night temperatures during the ripening period. Where the growth of sugar-cane is dependent upon natural rainfall the sugar content of the crop may vary widely in wet and dry years.

Under the most favourable ripening conditions, found in some parts of Queensland, Australia, one ton of sugar may be extracted from six or seven tons of cane. In other areas, such as Guyana, eleven or twelve tons of cane must be processed to obtain one ton of sugar. These differences persist despite intensive sugar-cane breeding and variety selection programmes aimed at increasing both cane yields and sucrose content.

Good juice quality (the concentration of sucrose in expressed cane juice) does not develop in the absence of appropriate environmental conditions. In a situation where climatic factors favour continuous vegetative growth even those varieties which possess the most favourable array of genes for sucrose production are unable to ripen. To do this they would need to respond differently to similar environmental conditions at different stages of their growth cycle. While this may be genetically feasible, cane breeders have made no significant advances in this direction so far. In Barbados, for example, spectacular improvements in sugar yields resulted from the introduction of new hybrid cane varieties between 1935 and 1960, but these improvements were due to higher tonnages of cane per acre rather than higher sucrose contents.[1] The juice qualities of modern commercial canes in 1960 were no better than those of the 'noble' canes (i.e. *Saccharum officinarum* varieties) B.H. 10(12) and B 726 in 1935.

CHEMICAL RIPENING

As the attainment of good juice quality is strongly dependent upon climatic factors which cannot themselves be easily or economically manipulated, chemical treatments which can override the natural interactions between genetic and environmental factors may prove useful in increasing sucrose accumulation. This approach has led to the concept of 'chemical ripening', that is the use of plant growth regulating chemicals to retard the vegetative growth of sugar-cane for several weeks prior to harvest, without seriously affecting its photosynthetic capacity. Although no chemical has yet been used to ripen sugar-cane on a commercial basis, several candidate ripeners have given encouraging results in greenhouse tests and in field trials.

The sequence of events between the reception of an environmental stimulus by the plant and the response of the plant in terms of growth may be divided into five broad steps *(Figure 1)*. Each step in this generalised pathway may be susceptible to modification by a chemical ripener.

The first step involves the physical interaction between the environment and the plant. For example, uptake of water by the roots is affected by the moisture content of the soil and by the permeability of the roots to water. This interaction may be modified by treating the roots with decenyl succinic acid, which increases their water permeability.[60] Similarly, the rate at which water is lost from the leaves by transpiration may be altered by chemical treatments which affect the opening and closing of the stomata.[61, 62] Since stem elongation in sugar-cane appears to be sensitive to moisture stress in the aerial parts of the plant,[47] the application of a chemical which would stimulate transpiration might induce ripening of sugar-cane growing in soil with a high moisture content.

The second step involves the synthesis of plant growth hormones. Obviously, treatments with chemicals which inhibit the synthesis of particular hormones will modify the environmental-genetic interaction. Thus treatments with the anti-gibberellins (2-chloroethyl)-trimethyl-ammonium chloride (chlormequat) or N-dimethylaminosuccinamic acid retard stem elongation in sugar-cane for about two weeks.[63] Similarly, chemicals which accelerate or retard the breakdown of growth hormones in the plant are likely to affect growth. These include phenolic acids such as caffeic and chlorogenic

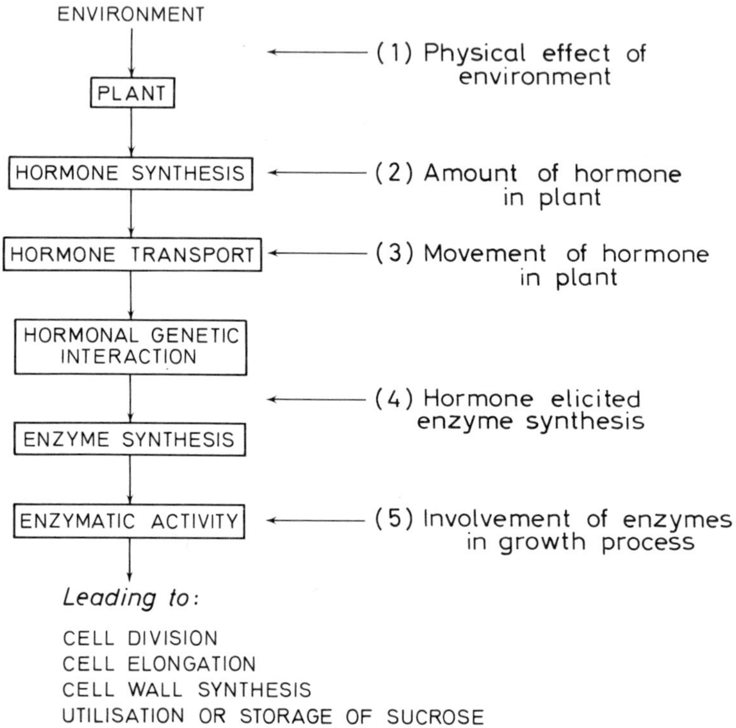

Figure 1. Diagram to illustrate the sequence of events between the reception of an environmental stimulus by the sugar-cane plant and the response of the plant in terms of growth

acid which inhibit IAA oxidase activity, and *p*-hydroxybenzoic acid and *p*-coumaric acid which enhance the oxidation of IAA in plant tissues.[64]

The natural synthesis of growth hormones in the plant may be augmented by supplying the plant with additional hormone. However, as we have seen, the natural hormonal balance in the plant is complex, involving several types of growth hormone, and appears to be rather delicately poised.

The application of a hormone to the plant may upset this balance drastically.

If the concepts of hormonal action presented earlier in this paper are correct, we may envisage the following situations. In immature stem apical tissues of fast growing sugar-cane there is sufficient auxin to permit the response of the tissues to gibberellins and cytokinins, but insufficient to promote the synthesis of ethylene. There is also sufficient gibberellin and cytokinin to promote cell elongation and cell division and to counteract premature ageing of the immature cells in response to ABA, but insufficient to prevent the progressive maturation and senescence of older cells. There is sufficient ABA to promote these last processes. In slow growing sugar-cane the supply of cytokinins and gibberellins to the meristematic tissues is restricted so that cell division and cell elongation tend to cease, or the synthesis of ABA is increased and tends to counteract cytokinin and gibberellin activity. There is an accumulation of auxin in the tissues which promotes ethylene synthesis and hence its own destruction. Low levels of gibberellin and cytokinin permit a heightened response of the tissues to ABA resulting in the premature maturation of cells. In both of these situations carbohydrate is incorporated into structural materials, albeit at different rates.

The ideal ripener would suspend all growth processes, cell division, cell elongation and differentiation, and the formation of new cell wall material. All tissues would be maintained in an unchanged condition and all available carbohydrate would move into storage. No doubt there is a particular hormonal balance which could produce this effect, although the balance would vary in different tissues. The hormonal balance appropriate to such 'suspended' growth would appear to be one at which there is just enough gibberellin and cytokinin to cancel out the effect of ABA, with sufficient auxin to maintain apical dominance (i.e. prevent the development of lateral buds) and possibly to retard root growth, and enough ethylene to maintain a favourable balance between auxin synthesis and auxin destruction *(Table 1)*. However, even if this balance could be achieved it would be difficult to maintain for any length of time against the natural tendency of the plant to re-establish its natural response to the environment. In practice, therefore, an effective ripener will achieve something less than this ideal, perhaps causing a gradual progression from a state of rapid vegetative growth, through a phase of suspended growth to one of slow growth.

In the early work on cane ripening, much attention was given to the possible use of the synthetic auxin 2, 4-dichlorophenoxyacetic acid (2, 4-D) as a chemical ripener. Some authors reported increases in the sucrose content of sugar-cane following treatment with 2, 4-D applied as a foliar spray,[65, 66] but others were unable to obtain a similar response

Table 1. HYPOTHETICAL STATES OF THE HORMONAL BALANCE IN IMMATURE STEM APICAL TISSUE OF SUGAR-CANE GROWING AT DIFFERENT RATES

Rate of Growth	*Relative levels of plant growth hormones*				
	Auxin	*Gibberellin*	*Cytokinin*	*ABA*	*Ethylene*
Fast	medium	high	high	low	low
Slow	high, becoming low	low	low	high	high, becoming low
'Suspended'	medium-high	medium	medium	medium	medium

of sugar-cane to this chemical.[67–70] These variable results may possibly be attributable to local differences in environmental conditions. Allen[71] found that sub-lethal applications of 2, 4-D to cowpea (*Vigna unguiculata* (L.) Walp.) plants increased leaf phosphofructokinase (PFK) activity in the light, but decreased the activity of this enzyme in the dark. At a daily photoperiod of approximately 12 hours and a more-or-less constant temperature of 78°F (25·6°C) the average PFK activity of leaf tissue treated with 2, 4-D was decreased by 11·4% over 24 h. PFK is an important regulator of glycolytic respiration in plants,[72] and a reduction in its activity probably indicates a lowered rate of respiration. If 2, 4-D has a similar effect on sugar-cane tissues then photoperiod and temperature could influence its effectiveness as a ripener. Best results might be expected under short day conditions with warm nights.

Another synthetic auxin, 2, 3, 6-trichlorobenzoic acid (TBA) has been found to induce ripening in sugar-cane.[73, 74] Vlitos and Lawrie[74] also found that 'Pesco 1815', a 1:3 mixture of TBA with 2-methyl-4-chlorophenoxyacetic acid (MCPA), was an effective ripener, although other workers have reported negative results with this chemical.[75, 76]

Yates[63] has made a more careful study of the effect of Pesco 1815 on the growth of sugar-cane under greenhouse conditions. He found that treatment with this chemical caused an initial acceleration of stem elongation, followed by a prolonged suppression of growth. The initial growth response was largely confined to one immature internode, and the growth of the younger internodes was completely suppressed. Similar results were obtained with applications of MCPA or TBA alone.[77] Senescence of the oldest leaves was accelerated and the emergence of new leaves was prevented by treatment with Pesco 1815, and in some experiments this chemical proved phytotoxic to sugar-cane.

The growth response of sugar-cane to Pesco 1815 may be interpreted in terms of its possible effect upon the hormonal balance in the plant. In rapidly elongating immature stem tissue the relative levels of the five types of growth hormone are considered to be: auxin medium, gibberellin and cytokinin both high, ABA and ethylene both low *(Table 1)*. Treatment with Pesco 1815 would raise the auxin level, and as the synthetic auxins are not readily degraded by the plant, this increased level would tend to persist. Initially the higher level of auxin would tend to enhance the activities of the gibberellins and cytokinins, thus promoting cell elongation and cell division. Soon afterwards the higher level of auxin would tend to promote the synthesis of ethylene in the tissues. This would lead to the destruction of auxin, particularly that native to the plant, and the oxidation of metabolites and enzymes essential for growth. The growth promoting activities of gibberellins and cytokinins would be negated, and growth would be suspended. Premature senescence of the oldest leaves may be due to an enhanced activity or synthesis of ABA, leading under certain circumstances to a phytotoxic effect. Apart from its effect upon leaf tissues, Pesco 1815 appears to fulfill the requirements for a successful chemical ripener. The co-administration of a gibberellin or a cytokinin to the plants might possibly overcome this effect, by balancing the increased activity of ABA.

The effect of gibberellic acid (GA) on the accumulation of sucrose by sugar-cane has also been studied. In field experiments Villareal and Santos[78] and Coleman *et al.*[79] found no significant effect of GA on sugar yields, and Bates[80] reported that GA treatment caused a general reduction in juice quality. Later field trials in Hawaii[81] and Queensland[75] indicated that applications of GA may increase sucrose yields, particularly in areas with relatively low (subtropical) temperatures.

In greenhouse experiments, in which sugar-cane plants were kept at controlled temperatures, Bull[82] found that weekly applications of GA over a seven to eight weeks period increased the length, fresh weight, sugar and fibre contents of the stalks, but decreased leaf area and weight. The response of young (three months) plants was greatest at low temperature (17°C), but older (six months) plants gave their largest response at a higher temperature (35°C). In spite of the reduction in leaf area sugar yields per stalk were increased by GA treatment.

An increase in the dry weight of the shoots of grasses treated with GA was found to be correlated with decreased root growth,[83] and Bull[82] suggests that this hormone may cause a selective partitioning of assimilates into storage tissue. However, Blesa and Recalde[84] report that treatment of germinating sugar-cane setts with GA increased root growth.

In terms of its possible effect upon the hormonal balance, treatment of sugar-cane with GA would be expected to stimulate stem elongation, particularly when the level of auxin is fairly high (i.e. in slow growing cane). This hormone might also tend to retard senescence by opposing the activity of ABA. If this is so, the observed reduction in leaf area due to GA may be compensated to some extent by a delayed senescence of older leaves. It is known that GA does not increase photosynthetic activity per unit leaf area.[85]

Preliminary experiments with ABA and with 'Ethrel' (2-chloroethane phosphonic acid), a compound which releases ethylene in plant tissues, indicate that these chemicals may have potential as cane ripeners.[63] Both retard stem elongation for about two weeks and this is followed by a period of rapid 'compensatory' growth. As might be expected, ABA treatment hastens senescence of the older leaves.

Treatment with Ethrel causes the loss of apical dominance as evidenced by the development of side shoots, tillers and strut roots. This is consistent with the finding that ethylene stimulates peroxidase activity in plant tissues,[52] leading to the destruction of auxin.

Hormone transport is obviously an important step in the response of plants to environmental stimuli. There is evidence that most hormones are translocated in the plant as free acids (auxins) or readily water soluble conjugates (gibberellins and cytokinins). Several chemicals are known to inhibit auxin transport.[86] One of these, 2, 3, 5-tri-iodobenzoic acid, has been tested on sugar-cane but was found to be a weak growth retardant.[63]

The last two steps in the pathway between environmental stimulus and growth response involve the synthesis and activity of enzymes. Chemicals which suppress the synthesis or activity of enzymes involved in the utilisation of sucrose may prove useful as ripeners. Thus Alexander[87] found that treatment of sugar-cane with 6-azauracil, a pyrimidine analogue which may inhibit RNA synthesis, caused a decline in growth and invertase activity, accompanied by an increase in sucrose content. The leaves of the plants were not affected, but the chemical caused malformation or death of meristematic tissue at the stem apex. Alexander[87, 88] also reports increased sucrose accumulation as a result of treating sugar-cane with molybdenum or tungsten, which inhibit phosphatase activity, and with silicon, which inhibits invertase activity. It is interesting to note that co-administration of GA with 6-azauracil or with silicon appears to enhance their ripening effect.

REFERENCES

1. STEVENSON, G. C., *Genetics and Breeding of Sugar Cane,* 1st edn, Longmans, Green & Co. Ltd, London (1965)
2. CROCKER, W. C., HITCHCOCK, A. E. and ZIMMERMAN, P. W., 'Similarities in the effects of ethylene and the plant auxins', *Contr. Boyce Thompson Inst. Pl. Res.,* **7,** 231–248 (1935)
3. MAPSON, L. W. and ROBINSON, J. E., 'The relation between oxygen tension, biosynthesis of ethylene, respiration and ripening changes in banana fruit', *J. Fd Technol.,* **1,** 215–225 (1966)
4. BURG, S. P. and BURG, E. A., 'Ethylene action and the ripening of fruits', *Science, N.Y.,* **148,** 1190–1196 (1965)
5. OSBORNE, D. J., 'Ethylene as a plant hormone', *Soc. chem. Ind. Monogr.,* **31,** 236–250 (1968)
6. BURG, S. P. and BURG, E. A., 'The interaction between auxin and ethylene and its role in plant growth', *Proc. natn. Acad. Sci., U.S.A.,* **55,** 262–269 (1966)
7. VAN OVERBEEK, J., 'Plant hormones and regulators', *Science, N.Y.,* **152,** 721–731 (1966)
8. KORNER, A., 'Regulation of the rate of synthesis of messenger ribonucleic acid by growth hormone', *Biochem. J.,* **92,** 449–456 (1964)
9. KEY, J. L., 'Ribonucleic acid and protein synthesis as essential processes for cell elongation', *Pl. Physiol.,* **39,** 365–370 (1964)

10. MASUDA, Y. and YANAGISHIMA, N., 'RNA functional in auxin action of expanding tuber tissue of Jerusalem artichoke', *Pl. Cell Physiol., Tokyo,* **5,** 365–368 (1964)
11. CHANDRA, R. G. and VARNER, J. E., 'Gibberellic acid-controlled metabolism of RNA in aleurone cells of barley', *Biochim. biophys. Acta,* **108,** 583–592 (1965)
12. TATA, J. R., In: *Mechanisms of Hormone Action,* Ed. by P. Karlson, Verlag, Stuttgart, 173 (1965)
13. BEERMAN, W. and CLEVER, U., 'Chromosome puffs', *Scient. Am.,* **210,** 50–58 (1964)
14. NOODÉN, L. D. and THIMANN, K. V., 'Inhibition of protein synthesis and of auxin-induced growth by chloramphenicol', *Pl. Physiol.,* **40,** 193–201 (1965)
15. NOODÉN, L. D. and THIMANN, K. V., 'Action of inhibitors of RNA and protein synthesis on cell enlargement', *Pl. Physiol.,* **41,** 157–164 (1966)
16. BONNER, J., *The Molecular Biology of Development,* 1st edn, Clarendon Press, Oxford (1965)
17. GAYLER, K. R. and GLASZIOU, K. T., 'Plant enzyme synthesis: hormonal regulation of invertase and peroxidase synthesis in sugar cane', *Planta,* **84,** 185–194 (1969)
18. BRANDES, E. W. and VAN OVERBEEK, J., 'Auxin relations in hot water treated sugar cane stems', *J. agric. Res.,* **77,** 223–238 (1948)
19. CUTLER, H. G. and VLITOS, A. J., 'The natural auxins of the sugar cane II. Acidic, basic and neutral growth substances in roots and shoots from twelve days after germination of vegetative buds to maturity', *Physiologia Pl.,* **15,** 27–42 (1962)
20. VLITOS, A. J. and CUTLER, H. G., 'The natural auxins of the sugar cane I. A paper chromatographic separation of the growth factors present in true seed', *Proc. Br. W. Indies Sug. Technol.,* **1960,** 113–127 (1961)
21. MOST, B. H. and VLITOS, A. J., 'Preliminary studies of gibberellin-like substances in sugarcane', In: *Régulateurs Naturels de la Croissance Végétale,* Ed. by J. P. Nitsch, C.N.R.S., Paris, 287–302 (1964)
22. MOST, B. H. and VLITOS, A. J., 'Gibberellins of sugar cane', *Pl. Physiol.,* **41,** 1090–1094 (1966)
23. MOST, B. H. and VLITOS, A. J., 'Studies on endogenous gibberellin-like substances in sugar cane', *Proc. int. Soc. Cane Technol.,* **12,** 610–624 (1967)
24. MOST, B. H. and VLITOS, A. J., 'The diffusible gibberellins from the stem apex and germinating lateral buds of sugarcane', In: *Transport of Plant Hormones,* Ed. by Y. Vardar, North-Holland Publishing Co., Amsterdam, 380–392 (1968)
25. MOST, B. H., 'A sugarcane spindle bioassay for gibberellins and its use in detecting diffusible gibberellins from sugarcane', In: *Biochemistry and Physiology of Plant Growth Substances,* Ed. by F. Wightman and G. Setterfield, The Runge Press Ltd, Ottawa, 1619–1633 (1968)
26. MOST, B. H., 'Studies of the gibberellin-like substances from sugarcane', *A. Rep. Tate & Lyle Ltd Res. Centre,* **1967,** 243–255 (1968)
27. BULL, T. A. and O'TOOLE, M. B., 'The gibberellin-like compounds of sugarcane', *A. Rep. Tate & Lyle cent. agric. Res. Stn, Trinidad,* **1966,** 225–236 (1967)
28. MOST, B. H., 'Diffusible and bound gibberellin-like substances in sugarcane', *A. Rep. Tate & Lyle cent. agric. Res. Stn, Trinidad,* **1966,** 257–262 (1967)

29. MOST, B. H., 'Growth hormone studies', *A. Rep. Tate & Lyle Ltd Res. Centre,* **1968,** 27–29 (1969)
30. MOST, B. H., 'Growth inhibitors of sugarcane I.' (In preparation)
31. THIMANN, K. V., 'Plant growth substances; past, present and future', *A. Rev. Pl. Physiol.,* **14,** 1–18 (1963)
32. SKOOG, F. and MILLER, C. O., 'Chemical regulation of growth and organ formation in plant tissues cultured *in vitro*', *Symp. Soc. exp. Biol.,* **11,** 118–131 (1957)
33. WICKSON, M. and THIMANN, K. V., 'The antagonism of auxin and kinetin in apical dominance', *Physiologia Pl.* **11,** 62–74 (1958)
34. ASPINALL, D., PALEG, L. G. and ADDICOTT, F. T., 'Abscisin II and some hormone-regulated plant responses', *Aust. J. biol. Sci.,* **20,** 869–882 (1967)
35. CHRISPEELS, M. J. and VARNER, J. E., 'Hormonal control of enzyme synthesis: on the mode of action of gibberellic acid and abscisin in aleurone layers of barley', *Pl. Physiol.,* **42,** 1008–1016 (1967)
36. VAN OVERBEEK, J., LOEFFLER, J. E. and MASON, M. I. R., 'Dormin (Abscisin II): inhibitor of plant DNA synthesis', *Science, N.Y.,* **156,** 1497–1499 (1967)
37. SANKHLA, N. and SANKHLA, D., 'Reversal of (±)-abscisin II induced inhibition of lettuce seed germination and seedling growth by kinetin', *Physiologia Pl.,* **21,** 190–195 (1968)
38. THOMAS, T. H., WAREING, P. F. and ROBINSON, P. M., 'Action of 'dormin' as a gibberellin antagonist', *Nature, Lond.,* **205,** 1270–1272 (1965)
39. MOST, B. H., 'Ripening of sugar cane. A model of how environmental changes can induce the ripening process in sugar cane', Tate & Lyle Ltd Res. Centre Seminar, private circulation (1969)
40. PHILLIPS, I. D. J. and JONES, R. L., 'Gibberellin-like activity in bleeding sap of root systems of *Helianthus annuus* detected by a new dwarf pea epicotyl assay and other methods', *Planta,* **63**, 269–278 (1964)
41. SKENE, K. G. M. and KERRIDGE, G. H., 'Effect of root temperature on cytokinin activity in root exudate of *Vitis vinifera* L.', *Pl. Physiol.,* **42**, 1131–1139 (1967)
42. CARR, D. J., REID, D. M. and SKENE, K. G. M., 'The supply of gibberellins from the root to the shoot', *Planta*, **63**, 382–392 (1964)
43. LUCKWILL, L. C. and WHYTE, P., 'Hormones in the xylem sap of apple trees', *Soc. chem. Ind. Monogr.*, **31**, 87–101 (1968)
44. VAADIA, Y. and ITAI, C., 'Interrelationships of growth with reference to the distribution of growth substances', In: *Root Growth*, Ed. by W. J. Whittington, Butterworths, London, 65–79 (1969)
45. WRIGHT, S. T. C., 'An increase in the "inhibitor β" content of detached wheat leaves following a period of wilting', *Planta*, **86,** 10–20 (1969)
46. MOST, B. H., 'The effect of moisture stress on the growth inhibitor content of sugar cane leaves', *Q. Rep. Tate & Lyle Ltd Res. Centre*, **3**, 131–133 private circulation (1969)
47. HUDSON, J. C., 'The available water capacity of Barbados soils', *Expl Agric.*, **5**, 167–182 (1969)
48. BURG, S. P. and BURG, E. A., 'Ethylene formation in pea seedlings; its relation to the inhibition of bud growth caused by indole-3-acetic acid', *Pl. Physiol.*, **43**, 1069–1074 (1968)
49. VAN OVERBEEK, J., 'Auxins', *Bot. Rev.,* **25**, 269–350 (1959)
50. NG, E. K. and AUDUS, L. J., 'Growth-regulator interactions in the growth of the shoot system of *Avena sativa* seedlings I. The growth of the first internode', *J. exp. Bot.,* **15**, 67–95 (1964)
51. SIEGEL, S. M., *The Plant Cell Wall,* 1st edn, Pergamon Press, Oxford (1962)

52. HALL, W. C. and MORGAN, P. W., 'Auxin-ethylene interrelationships', In: *Régulateurs de la Croissance Végétale*, Ed. by J. P. Nitsch, C.N.R.S., Paris, 727–745 (1964)
53. WALTON, D. C. and SONDHEIMER, E., 'Effects of abscisin II on phenylalanine ammonia-lyase activity in excised bean axes', *Pl. Physiol.,* **43**, 467–469 (1968)
54. HARTT, C. E., KORTSCHAK, H. P., FORBES, A. J. and BURR, G. O., 'Translocation of C^{14} in sugarcane', *Pl. Physiol.,* **38**, 305–318 (1963)
55. SACHER, J. A., HATCH, M. D. and GLASZIOU, K. T., 'Sugar accumulation in sugar cane III. Physical and metabolic aspects of cycle in immature storage tissue', *Pl. Physiol.,* **38,** 348–354 (1963)
56. SACHER, J. A., HATCH, M. D. and GLASZIOU, K. T., 'Regulation of invertase synthesis in sugar cane by an auxin- and sugar-mediated control system', *Physiologia Pl.,* **16,** 836–842 (1963)
57. HATCH, M. D. and GLASZIOU, K. T., 'Sugar accumulation cycle in sugar cane II. Relationships of invertase activity to sugar content and growth rate in storage tissue of plants grown in controlled environments', *Pl. Physiol.,* **38**, 344–348 (1963)
58. GLASZIOU, K. T. and BULL, T. A., 'The relationship between total invertase activity and internode expansion in sugar cane stalks', *Proc. int. Soc. Sug. Cane Technol.,* **12**, 575–581 (1967)
59. HAWKER, J. S. and HATCH, M. D., 'Mechanism of sugar storage by mature stem tissue of sugarcane', *Physiologia Pl.,* **18**, 444–453 (1965)
60. KUIPER, P. J. C., 'Inducing resistance to freezing and desiccation in plants by decenylsuccinic acid', *Science, N.Y.,* **146**, 544–546 (1964)
61. ZELITCH, I., 'Biochemical control of stomatal opening in leaves', *Proc. natn. Acad. Sci. U.S.A.,* **47**, 1423–1433 (1961)
62. WAGGONER, P. E., MONTEITH, J. L. and SZEICZ, G., 'Decreasing transpiration of field plants by chemical closure of stomata', *Nature, Lond.,* **201**, 97–98 (1964)
63. YATES, R. A., 'Chemical ripening', *A. Rep. Tate & Lyle Ltd Res. Centre,* **1968,** 21–26 (1969)
64. ZENK, M. H. and MÜLLER, G., '*In vivo* destruction of exogenously applied indoleacetic acid as influenced by naturally occurring phenolic acids', *Nature, Lond.,* **200**, 761–763 (1963)
65. BEAUCHAMP, C. E., 'Effects of 2, 4-D on sugar content of sugar cane', *Sugar J.,* **13** (5), 57–70 (1950)
66. CHACRAVARTI, A. S., SRIVASTAVA, D. P. and KHANNA, K. L., 'Application of phytohormone to sugarcane', *Proc. int. Soc. Sug. Cane Technol.,* **9**, 355–364 (1956)
67. LOUSTALOT, A. J., CRUZADO, H. J. and MUZIK, T. J., 'The effect of 2, 4-D on sugar content of sugarcane', *Sugar J.,* **13** (5), 78 (1950)
68. HASKEW, H. C., 'Effect on C.C.S. of spraying 2, 4-D on to growing cane', *Cane Grow. Q. Bull., Queensland,* **17**, 52–53 (1953)
69. LUGO-LOPEZ, M. A., SAMUELS, G. and GRANT, R., 'Failure of preharvest sprays with 2, 4-D and maleic hydrazide to affect the sucrose content of sugar cane', *J. Agric. Univ. P. Rico,* **37,** 44-51 (1953)
70. YATES, R. A. and BATES, J. F., 'Preliminary experiments on the effects of chemicals on the ripening of sugar cane', *Proc. Br. W. Indies Sug. Technol.,* **1957**, 174–189 (1958)
71. ALLEN, T. J., *The effect of 2, 4-dichlorophenoxyacetic acid on the activity of phosphofructokinase in Vigna unguiculata (L.) Walp. leaf tissue*, Ph.D. dissertation, Texas A. & M. University, U.S.A. (1969)

72. DENNIS, D. T. and COULTATE, T. P., 'Phosphofructokinase, a regulator enzyme in plants', *Biochem. biophys. Res. Commun.,* **25**, 187–191 (1967)
73. NICKELL, L. G. and TANIMOTO, T., 'Effects of chemicals on ripening of sugar cane', *Rep. a. conf. Hawaii. Sug. Technol.,* **24**, 152–166 (1965)
74. VLITOS, A. J. and LAWRIE, I. D., 'Chemical ripening of sugar cane. A review of field studies carried out in Trinidad over a five year period', *Proc. int. Soc. Sug. Cane Technol.,* **12**, 429–445 (1967)
75. GLASZIOU, K. T., 'Chemical control of growth and ripening', *Rep. David North Plant Res. Centre,* 48–55 (1964)
76. ANONYMOUS, 'The influence of chemicals on juice quality', *A. Rep. Res. Dep. Sug. Mfrs. Ass. Jamaica,* 5 (1963)
77. YATES, R. A., 'Screening of candidate chemical ripeners', *Q. Rep. Tate & Lyle Ltd Res. Centre,* **2**, 132–144 (1969)
78. VILLAREAL, E. P. and SANTOS, O., 'The effect of gibrel upon the growth and yield of sugarcane', *Proc. a. Conv. Philipp. Sug. Technol.,* **6**, 87–91 (1958)
79. COLEMAN, R. E., TODD, E. H., STOKES, I. E. and COLEMAN, O. H., 'The effect of gibberellic acid on sugar cane', *Proc. int. Soc. Sug. Cane Technol.,* **10**, 588–603 (1960)
80. BATES, J. F., 'Preliminary experiments on the effects of gibberellic acid on germination, growth and flowering in sugarcane', *Proc. Br. W. Indies Sug. Technol.,* **1957**, 165–173 (1958)
81. TANIMOTO, T. and NICKELL, L. G., 'Re-evaluation of gibberellins for field use in Hawaii', *Rep. a. Conf. Hawaii. Sug. Technol.,* **25**, 184–190 (1967)
82. BULL, T. A., 'The effects of temperature, variety and age on the response of *Saccharum* spp. to applied gibberellic acid', *Aust. J. agric. Res.,* **15**, 77–84 (1964)
83. BRIAN, P. W., ELSON, G. W., HEMMING, H. G. and RADLEY, M., 'The plant-growth-promoting properties of gibberellic acid, a metabolic product of the fungus *Gibberella fujikuroi*', *J. Sci. Fd Agric.,* **5**, 602–612 (1954)
84. BLESA, C. A. and RECALDE, L., 'The effect of gibberellic acid on roots. I. The effect of a gibberellic acid-lanolin mixture on root development of sugar cane cuttings', *Phyton, B. Aires,* **16**, 183–188 (1961)
85. HABER, A. H. and TOLBERT, N. E., 'Photosynthesis in gibberellin-treated leaves', *Pl. Physiol.,* **32**, 152–153 (1957)
86. MCREADY, C. C., 'Translocation of growth regulators', *A. Rev. Pl. Physiol.,* **17**. 283–294 (1966)
87. ALEXANDER, A. G., 'The use of chemicals for sucrose control in sugar cane', *Sug. Azúc.,* **64**, 21–25 (1969)
88. ALEXANDER, A. G., 'Introduction of varying sugar levels in the leaves of immature sugar cane by use of acid phosphatase inhibitors', *J. Agric. Univ. P. Rico,* **49**, 35–59 (1965)

IV

THE ABSORPTION AND METABOLISM OF SUCROSE IN ANIMALS

Chapter Fourteen

THE ROLE OF THE INTESTINE IN SUCROSE ABSORPTION

W. F. WIDDAS
Professor of Physiology, Bedford College, University of London

1 SUCROSE AS A CARBOHYDRATE OF THE DIET

In nature the organic compounds in greatest abundance are carbohydrates of high molecular weight.

Cellulose, an important structural molecule in plants, is a compound of polymerised glucose molecules. The sugar residues are joined into long chains by glycosidic linkages but on chemical hydrolysis the simple hexose, glucose, is recovered. In cellulose, the glycosidic linkages are of the β form and are not subject to hydrolysis by the enzymes present in mammalian salivary and gastro-intestinal secretions.

Starch, which is a storage carbohydrate of plants, is also made up of glucose polymerised into chains. The glycosidic links are of the α form and can be hydrolysed by mammalian digestive enzymes. The end products of hydrolysis by amylase are mixtures of glucose, maltose and oligosaccharides (in which a few glucose residues are still glycosidically linked).

Glycogen, which is a storage carbohydrate found in the liver and muscles of animals, is chemically similar to starch and gives similar hydrolysis products.

The smallest compounds in which sugars are glycosidically linked are disaccharides, many examples of which exist in nature. The most important from a dietary point of view are:

Lactose (milk sugar), a disaccharide solely of mammalian origin in which a molecule of glucose is glycosidically linked to a molecule of galactose.

Maltose, a disaccharide composed of two molecules of glucose which, with free glucose, form the major end-products of the enzymatic hydrolysis of starch by amylase.

Sucrose, a disaccharide in which a molecule of glucose is glycosidically linked to a molecule of fructose. This disaccharide is peculiar in that the anomeric carbon atom of each hexose is involved in the glycosidic link and thus the molecule is chemically unreactive (will not reduce alkaline solutions of Cu^{++} ions) although hydrolysis (enzymatic or chemically catalysed) proceeds without difficulty.

Sucrose, apart from its domestic source (from sugar-cane and sugar-beet), occurs to a small extent in the juices of many fruits which also contain the free simple sugars: glucose (grape sugar) and fructose (fruit sugar).

Thus, although small quantities of dietary carbohydrates are in the form of free simple sugars, the major portions are polymerised forms ranging from disaccharides up to high molecular weight starches and cellulose. Sucrose with maltose and lactose comprise the simplest of the glycosidically-linked molecules of the diet.

2 ENZYMES FOR HYDROLYSING CARBOHYDRATES

As the simple sugars are the forms in which carbohydrates are absorbed into the blood stream, hydrolysis of dietary carbohydrates is an important digestive function and involves catalytic enzymes.

SALIVARY AND PANCREATIC AMYLASE

An enzyme that can hydrolyse starch (amylase) is found in salivary and pancreatic secretions. This enzyme splits α, 1, 4-glucosidic bonds, except for that of maltose, in a random manner: hence, the form of starch composed of simple long chains (amylose) is broken down into a mixture of glucose and maltose. Glycogen and the more branched form of starch (amylopectin) contains α, 1, 6-glucosidic bonds at the branch points and these are not split by amylase: hence, the products of hydrolysis of amylopectin (and glycogen) include isomaltose and oligosaccharides in which α, 1, 6-glucosidic bonds abound.

Thus, even without disaccharides in the diet, enzymes capable of hydrolysing maltose (α, 1-4 bond) and isomaltose

(α, 1-6 bond) would be essential to complete the digestion of starch, the main dietary carbohydrate of adults.

DISACCHARIDASES

Dahlqvist[1] investigated the distribution of carbohydrate-splitting enzymes along the digestive tract of the pig. The stomach has only a weak amylase activity (possibly from saliva) and the colon shows much weaker activity than the small intestine. Whereas pancreatic juice, which flows into the duodenum, possesses amylase activity, it has only weak maltase activity, but enzymes which split the glycosidic bond in disaccharides are abundant in the mucosa of the small intestine.

Dahlqvist and Borgström[2] investigated the rates of absorption and degree of hydrolysis in the lumen of the human small intestine at different levels. The results indicated that lactose is acted on in the duodenum and jejunum, maltose chiefly in the jejunum and proximal ileum, and sucrose in the lower jejunum and ileum. Evidence for the presence of sucrase activity extended from the duodenum to terminal ileum, as did maltase activity, but there was little lactase activity in the lower ileum.

In contrast to the amylase which is secreted in pancreatic juice and presumably acts in the intestinal luminal fluid, the disaccharidases are associated with the intestinal mucosa and in homogenised extracts are bound to particle fractions.

In man, it was found that particle-bound disaccharidases could be solubilised by papain digestion or by 'autolysis'.[3] Separation by Sephadex G-200 shows several fractions with maltase activity (α-disaccharidases); a large fraction hydrolyses both sucrose (sucrase) and isomaltose (isomaltase) but other enzyme fractions hydrolyse maltose but not sucrose or isomaltose. The sucrase enzymes show peak activity at pH 5·8, although divisible chromatographically into sucrase 1 and sucrase 2 presumably by molecular size.

Although one fraction contains both sucrase and isomaltase activity, Auricchio and his colleagues[4] consider that separate enzymes functions are involved. Thus there are differences in the time course of heat inactivation at 50°C. Also isomaltose, when present with sucrose, has no effect on the rate of hydrolysis of the latter by the sucrase enzyme, nor has sucrose an inhibitory effect on isomaltase activity. The authors now recognise five maltase fractions: maltases 1 and 2 which have minimal sucrase or isomaltase activity (but may have some amylase activity) and which are inactivated only at higher temperatures; and maltases 3, 4 and 5 which are equivalent to sucrases 1 and 2 and isomaltase respectively.

The sucrase enzymes have a K_s for sucrose of 0·02 M and are found to be activated by Na^+ ions (NaCl 0·05 M) by some 30–40%. The same properties are shown by the particle-bound sucrases.

Two lactases (β-disaccharidases) are also observed in extracts of jejunum and ileum, as well as carbohydrase enzymes, the physiological significance of which is unknown (e.g. trehalase).

The intestinal mucosa is also a site of high alkaline phosphatase activity, and histochemically this can be shown to be associated with the 'brush border' of the mucosal cells. Homogenisation in 0·005 M EDTA completely disrupts the mucosal cells with the exception of the 'brush borders' which can be separated by differential centrifugation.[5] Sucrase and maltase activity is associated with these 'brush borders'. Not only are disaccharidases present but also alkaline phosphatase and leucyl naphthylamidase (two enzymes which can be localised histochemically) in proportions to suggest that the 'brush border' was the only important localisation of the intestinal disaccharidases.[6] The 'brush border' or absorptive pole of the mucosal cell shows, under the electron microscope, many finger-like processes termed microvilli, and these are schematically represented in *Figures 1 and 3*.[7] The localisation of disaccharidases along the microvilli has been proposed on the basis of electron microscope studies.[8]

Thus it is now established that the disaccharidases that split maltose, isomaltose and oligosaccharides (resulting from the digestion of starch by amylase) into glucose, and that also split dietary disaccharides (lactose and sucrose) into their component simple sugars, are closely correlated with the intestinal absorption surface.

3 ALIMENTARY SITE OF CARBOHYDRATE ABSORPTION

There is considerable evidence that the stomach and colon are not involved and that normal carbohydrate absorption occurs chiefly in the small intestine.[9] The capacity for absorption far exceeds the normal dietary load and although little absorption occurs in the stomach, gastric secretions tend to dilute administered solutions, especially when hypertonic. These factors tend to confuse interpretations of results based on intragastric administration of sugars but Cori,[10] using such methods, showed important differences in the rates of absorption of different monosaccharides which indicated that special mechanisms aided hexose absorption.

Glucose and galactose (two of the sugars most rapidly absorbed) are also absorbed from the lumen of the intestine when the concentration is lower than that in the blood. Such transfer from the mucosal to the serous side of the intestinal wall, against the diffusion gradient, is recognised as active transfer and, although supported by earlier work on whole animals, its intensive study only became possible with *in vitro* techniques.[11] Rat small intestine *in vitro* can transport glucose from mucosal to serosal fluid against a considerable concentration gradient.[12, 13] The active transport is oxygen dependent and inhibited by 0·1 per cent phlorizin. The introduction of the everted sac of small intestine,[14] which ensures adequate oxygenation of the mucosal surface while, incidentally, containing the serosal fluid to a relatively small volume, has been the basis of practically all subsequent *in vitro* work.

Measurements of glucose uptake[15, 16] suggested that sugar was entering the mucosal cells by a membrane 'carrier' process analogous to that studied in erythrocytes.[17]. This analogy received very strong confirmation by the demonstration of uphill transfer by counterflow,[18] a phenomenon predictable from mobile carrier transfers[19] and subsequently shown to occur in rabbit erythrocytes[20] and human erythrocytes.[21] However, accumulation against the gradient (which is not seen in erythrocytes) also occurs across the luminal-facing membrane or 'brush border'. The concentration of D-galactose (in frozen dried sections) is greatest within the mucosal cell[22] and autoradiographic and other evidence[7, 23] confirms this view. Thus, not only is the mucosal cell 'brush border' the site of absorption of simple sugars but also the site of a mechanism capable, in the case of actively transferred sugars, of effecting such absorption against a concentration gradient. Since the 'brush borders' are also the sites of the disaccharidases they may be identified as the structures which endow carbohydrate absorption with its well defined characteristics. Schematic representations of this region at increasing magnification are given in the *Figures* in the order *1*, *3* and *2*.[7]

4 CHARACTERISTICS OF INTESTINAL ABSORPTION OF SIMPLE SUGARS

Investigations (chiefly with everted sacs) have suggested a broad division among simple sugars into those which can be actively absorbed and those passively absorbed. The division is based on the demonstration of accumulation against the gradient and does not necessarily imply different mechanisms. However, the two groups have sufficiently different properties to merit their separate consideration.

(*a*) ACTIVELY ABSORBED SUGARS

Glucose and galactose and a number of deoxy-derivatives are transferred against a concentration gradient, as also are D-allose and D-glucoheptulose.[24] In the deoxy-derivatives it is essential to retain a hydroxyl group at C2,[25] but this group and the pyranose ring common to glucose and galactose are the only specificity features required.

Anaerobic conditions prevent active uptake and the situation is complicated by the absorbing cells deriving energy from glucose metabolism. The presence of organic phosphates derived from glucose could thus either have a metabolic significance or could be a step in the active accumulation. That phosphorylation was not involved in accumulation *per se* was shown by adding ^{14}C-galactose to the medium while the everted sac was actively absorbing glucose.[26] On isolating glucose-6-phosphate from the cells and glucose from the serosal fluid, it was found that, compared with glucose-6-phosphate, very little of the transferred glucose was labelled by the ^{14}C-galactose; much less than would have been expected if the accumulated glucose went through the glucose-6-phosphate pool.

New light on the nature of the active absorption has come from observations that glucose and galactose cause a rapid change in the transepithelial electrical potential difference, the mucosal side becoming 5–9 mv more negative. When the potential is held steady, there is an increase in the short-circuit current which has been equated with the increase in transfer of sodium ions from mucosal to serosal surfaces. The relationship has been observed in a number of species e.g. Greek tortoise,[27] rat small intestine[28] and rabbit ileum.[29] 3-*O*-methyl glucose, which is accumulated without being metabolised, also increases the short-circuit current, whereas

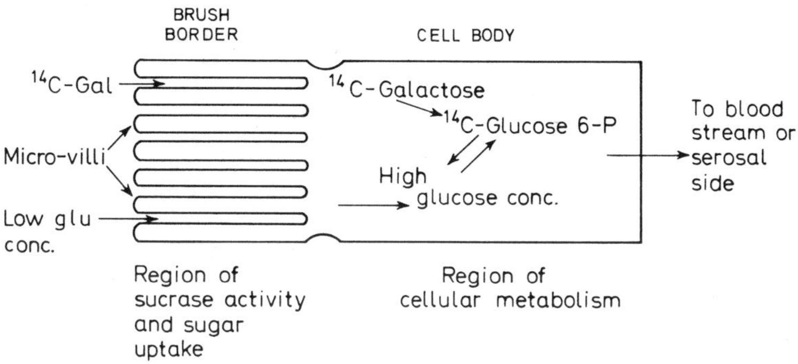

Figure 1. Schematic representation of an intestinal mucosal cell. The brush border faces the intestinal lumen and is the region of sucrase hydrolysis and absorption of the simple sugars. The simultaneous uptake of labelled galactose and unlabelled glucose was used by Landau and Wilson[26] to show that accumulated glucose was not phosphorylated (see text)

fructose and mannose, which are metabolised but not actively accumulated, do not affect the potential or short-circuit current. These findings exclude an indirect metabolic effect. Phlorizin, which inhibits accumulation of actively transferred sugars, also blocks the increase in short-circuit current.

The increase in short-circuit current is related to glucose concentration giving a non-linear saturable-type curve with a half-saturation value of about 4 mM at 38·5°C.[29] However, this applies only in the presence of Na^+. One effect of the absence of sodium is to alter the apparent half-saturation concentration for the accumulated sugar so that it appears as if sugar uptake involves a membrane carrier with Michaelis-Menten type kinetics, the K_m of which depends on the sodium concentration.[30]

The dependence of sugar accumulation by guinea pig intestine on the presence of sodium in the medium was observed by Riklis and Quastel,[31] and confirmed in the rat by Csáky and Zollicoffer.[32] Not only is the absence of Na^+ (as by Tris substitution) inhibitory, but other monovalent ions are also inhibitory. For instance, with increased $[K^+]$

there is a close parallel between the inhibition of sugar transport and of Na^+ influx.[33]

Mucosal cells were estimated to contain 140 mM[K^+] and 50 mM[Na^+] when incubated in a medium of 140 mM[Na^+] and 15 mM[K^+], and Na^+ was being transferred towards the serosal surface to maintain the steady state.[34] Flux measurements suggest that the inward leak of sodium across the brush border must be well in excess of net serosal transfer and that a major part must be ejected into the luminal fluid. This would agree with the presence in the 'brush border' of ATPase enzymes which function as a sodium pump.

However, there is considerable evidence that the extra Na^+ transfer associated with sugar accumulation does not result in increased 'brush border' sodium pumping. Thus, ouabain, which inhibits the sodium pump, has no effect on the current changes associated with sugar transport when on the mucosal side but is effective on the serosal side. Intracellular microelectrodes which show the interior of the mucosal cells to be about 20 mv negative to the mucosal surface and 23 mv negative to the serosal surface in the absence of substrate only show a change relative to the serosal side while substrate is being accumulated. Thus a stimulation of an electrogenic sodium pump on the serosal surface, presumed to occur due to the extra sodium ions entering across the 'brush border', must be the cause of the potential and short-circuit current changes.

The energy for the active accumulation of sugars can thus be seen to be derived from the flow of sodium ions along the downhill electrochemical gradient (for these ions) from mucosal surface across the 'brush border' into the cell interior. The serosal poles (and inter-cellular channels) are presumed to be sites for active sodium extrusion from the cells and, helped by extrusion across the brush border, the low internal sodium concentration is maintained. The sodium pumps directed towards the serosal surface, by responding to increased Na^+ load by greater activity, preserve the electrochemical gradient even during active accumulation of sugars. Thus the indirect but ultimate source of energy would be derived from ATP through the Na^+ and K^+ activated (ouabain-sensitive) membrane ATPase which is now recognised to be associated with Na^+ extrusion from and K^+ accumulation in nerves, red cells, and many other tissues.[35]

The coupling of sugar and Na^+ transfer across the brush border membrane is schematically represented in *Figure 2*.

The high cellular [K^+] competes for the carrier on the intracellular side of the brush border membranes and by displacing Na^+ lowers the affinity of the carrier for the sugar; sodium and glucose are enabled to enter the cell and potassium to leave it so that the sugar will be trapped and accumulate to

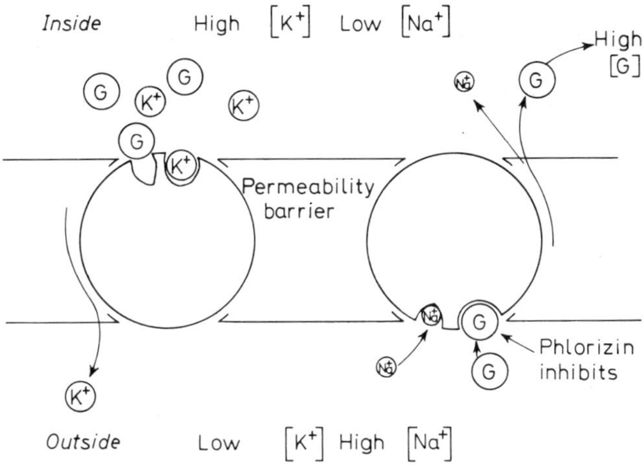

Figure 2. Schematic representation of the carrier mechanism for the sodium coupled transfer of glucose. The hexose site has high affinity when Na^+ *occupies the cation site but low affinity when* K^+ *occupies that site. The* Na^+ *and* K^+ *gradients across the permeability barrier thus produce an asymmetry in glucose (or galactose) transfer. (After Crane[7])*

high concentration as long as the sodium and potassium electro-chemical gradients are maintained. The free sugar in the cell may leave the serosal surface by diffusion or by a facilitated transfer not linked to sodium transfer.

The stoichiometry of the system is still uncertain and it should not be assumed that each sugar molecule accumulated allows only one extra Na^+ ion to enter. Some results suggest that as many as five Na^+ ions may accompany each glucose molecule accumulated, but these experiments may have been carried out under conditions of high saturation and less than optimal conditions.[36] (For a fuller discussion *see* Crane.[23])

(*b*) PASSIVELY ABSORBED SUGARS

Mannose, sorbose and fructose, as well as a number of pentose sugars and *l*-enantiomorphs of glucose and galactose, are not accumulated against a concentration gradient by *in vitro* intestinal preparations. They cross the permeability barrier of the cell but in a direction determined by the concentration gradient.

Although there is now evidence[37] that some of these sugars may enter the mucosal cells on the same carrier as used by glucose and galactose (without accumulation against a gradient) it appears that an alternative pathway exists. Sorbose and fructose appear to enter at similar rates, and anoxia and metabolic inhibitors have no effect.[38] Ions that stimulate glucose uptake have no effect on fructose transfer and phlorizin, which inhibits glucose uptake, is similarly without effect.

Fructose is not transported against its concentration gradient in everted sacs of hamster intestine and its rate of transfer, while similar to the other sugars in this group, is only about one-sixth that of glucose.[39] Conversely, with a high serosal concentration, fructose is transferred to the mucosal side at 4–5 times the rate of glucose. In this preparation fructose transfer is actually increased by oxygen lack.

These views that fructose crosses the intestinal mucosa passively are supported, for the human intestine, by the investigations of Holdsworth and Dawson.[40]

Nevertheless, mannose and fructose are phosphorylated by hexokinase enzymes in mucosal cells and enter metabolic chains with possible conversion to glucose or lactic acid. The conversion depends on the species. The rat converts 10% of the absorbed fructose to glucose, whereas in guinea-pig intestine the conversion is as high as 70%.[41]

In contrast to the actively accumulated sugars, neither mannose nor fructose induce any changes in potential or short-circuit current across the intestinal epithelium although fluid transfer, which is also enhanced during glucose accumulation, is stimulated by fructose. However, fluid transfer may depend on metabolism,[42] and this effect may thus be separable from fluid transfer induced by the extra Na^+ transfer of active accumulation.

5 THE INTESTINAL HANDLING OF DISACCHARIDES

Although many experiments have been carried out on the 'absorption' of disaccharides by noting their rate of disappearance from intestinal fluid, there is no convincing evidence of their absorption unchanged into the blood stream. Less than 0·2% of a 25 g test dose of lactose was excreted in the urine during the following five hours as compared with 20% of a similar dose of xylose.[43] In sprue, disaccharides are found in the urine and it seems absorption may be evidence of mucosal damage.[44]

The normal mode of handling is thus hydrolysis into the component simple sugars which could be absorbed in accordance with the principles outlined above, and the necessary enzymes (disaccharidases) are present in the intestinal mucosa closely associated with the absorption surface in the 'brush border'.

Hydrolysis in the lumen of the intestine with subsequent absorption would not, however, fit the observations. The rate of 'absorption' of sucrose and lactose by the dog small intestine proceeds faster than expected if hydrolysis occurs in the lumen.[45] There is a more rapid appearance of fructose in the serosal fluid of everted sacs of rat small intestine when sucrose is on the mucosal side than fructose itself.[46] The kinetic advantage is clearly shown by the fact that neither increased volume, which would dilute any hydrolysed product released into the luminal medium, nor the presence of glucose oxidase in sufficient amounts to remove any glucose so released, appreciably reduces the tissue fructose during sucrose 'absorption' or the tissue glucose during sucrose or maltose 'absorption'.[5]

These results are consistent with hydrolysis within the mucosal cell. However, if hydrolysis is intracellular, it must be external to the permeability barrier across which glucose is actively accumulated.[5, 47] Further evidence is derived from fluid transfer studies based on the effects of inhibitors of glucose accumulation.[42] Thus during maltose 'absorption' phlorizin will markedly reduce the appearance of glucose on the serosal surface or the fluid transfer associated with active glucose transfer. The quantity of maltose split is, however, not reduced but the glucose released by hydrolysis appears in the mucosal fluid.

Lack of Na^+ ions in the mucosal solution has a similar effect to phlorizin in reducing the glucose appearing on the serosal side, but again is without effect on the amount of maltose split.[48] The effect on glucose penetration, but not on the rate of splitting, is also seen with glucose-1-phosphate which may be presumed to be split by alkaline phosphatase, an enzyme histochemically localised to the 'brush border' of the mucosal cells. That the phosphate split off appears in the mucosal fluid and does not equilibrate with intracellular phosphate had been shown earlier.[49]

The above data in conjunction with electron microscopic localisation of the plasma membranes of the microvilli of the 'brush borders'[50] would require that the disaccharidases are situated along the microvilli,[51] as represented schematically in *Figure 3* and as suggested from the knobs indicated in recent electron micrographs.[8] The glucose hydrolysed from

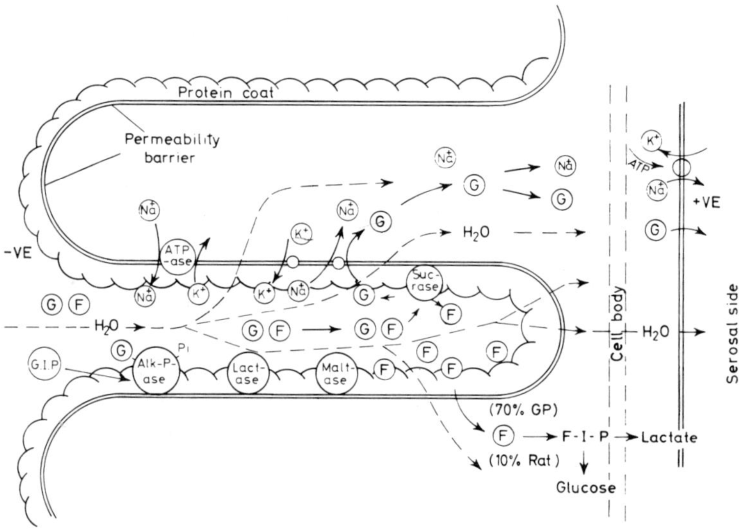

Figure 3. Schematic representation of sucrose digestion followed by absorption of the component sugars across the brush border. The disaccharidases are shown outside the permeability barrier but along the microvilli so that the hydrolysis products have a kinetic advantage over expectations based on their release into the intestinal lumen. The active uptake of glucose Ⓖ *exceeds the passive uptake of fructose* Ⓕ *so that more of the latter accumulates in the microvilli spaces and diffuses into the luminal fluid*

sucrose is released only in the α-form,[52] but this in itself conveys no kinetic advantage. It has thus been deduced that the sucrase enzyme must also be in close spatial relationship with the carrier that accumulates glucose and galactose, and with the alternative pathway used by fructose.[53]

Other possibilities considered were that sucrase may act as a permease catalysing the transfer of the glucose moiety on to the carrier or that the carrier may act as acceptor for a transglucosidase activity by the enzyme. Although sucrase is activated by Na^+, the possibility that it is also the glucose carrier is discounted both on kinetic grounds and from the properties of inhibitors. A possible factor in the kinetic advantage not usually given attention is the associated fluid movement also schematically indicated in *Figure 3.* Since actively absorbed glucose promotes extra Na^+ transfer and fluid transfer towards the serosal side, there should be a net bulk flow of luminal fluid along the narrow channels surrounding the microvilli. This would oppose diffusional loss of glucose and fructose from such channels towards the lumen. The osmotic withdrawl of water following the entry of Na^+ and glucose into the cell would tend to reduce the dimensions and fluid in such channels, and this should also help to increase the concentration of disaccharide hydrolysis products.

Maltose and lactose are hydrolysed to products which are actively absorbed and so the kinetic advantage from these various factors will work synergistically. Sucrose, on hydrolysis, yields equimolar quantities of glucose and fructose, and any kinetic advantage of the fructose over free fructose must either be associated with the active accumulation of glucose (e.g. fluid transfer) or must involve molecular participation of the sucrase enzyme in one of the ways suggested above.

In a study of sucrose 'absorption' by everted sacs of hamster intestine, it was shown chromatographically that, although glucose and fructose and some sucrose is detectable in the mucosal fluid at the end of one hour's incubation, glucose is the predominant sugar on the serosal side.[39] Thus, starting with sucrose (1000 mg/100 ml) on the mucosal side, 490 mg/100 ml glucose and 62 mg/100 ml fructose was found on the serosal side, while 268 mg/100 ml fructose and 103 mg/100 ml glucose was found on the mucosal side after one hour. The rates of hydrolysis of maltose and sucrose were estimated to be 2–3 times the maximal rate at which glucose could be absorbed, but, as previously mentioned, tests on the simple sugars showed that glucose was absorbed about six times as fast as fructose. These results are thus consistent with the 'brush border' hydrolysis of sucrose into its component sugars with subsequent absorption of simple sugars in which glucose is taken up more rapidly and with less loss by diffusion into the luminal fluid.

Such an interpretation on the handling of sucrose by the human intestine explains the results of comprehensive studies on volunteers with a double-lumen tube at different levels of the intestine.[54] A slow infusion of sugar solutions with a non-absorbable marker substance through the proximally opening tube was combined with sampling of the fluid in the neighbourhood of the distal orifice 15–30 cm further along the intestine. The advantage of the use of a non-absorbable marker in such studies was shown in experiments on the digestion and absorption of sucrose in the rat and in which the same features were observed.[47]

The results *in vivo* in the human were analogous to *in vitro* experiments.[39] Thus, in a typical result in which 66 mmol of sucrose were infused in 1 h, 39 mmol were hydrolysed; of the glucose released, 35 mmol were absorbed and 4 mmol were found intraluminally; of the fructose released, 21 mmol were absorbed and 18 mmol appeared in the luminal fluid. Thus an almost complete absorption of the glucose (liberated by sucrose hydrolysis) associated with a less efficient fructose absorption may be regarded as the typical characteristic of sucrose handling.

6 MALABSORPTION OF CARBOHYDRATES

Malabsorption syndromes have now been recognised in which either specific enzymes are absent or there is a specific deficiency in the active uptake of glucose and galactose (glucose–galactose malabsorption). These conditions have recently been reviewed[55] and will only be referred to here in so far as they lend confirmation to the physiological interpretations, given in previous sections, which concern sucrose.

Sucrose intolerance has been recognised as an inherited disease due to a congenital sucrase-isomaltase deficiency. Although the two enzymes have different inactivation properties and mutual competition between substrates do not occur, they are nevertheless inseparable and have been interpreted as a single building block of the brush border with two 'active sites'.[51] This also fits in with the suggestion that the disease is due to a single genetic factor.

The undigested disaccharides are not absorbed but pass on to the large intestine and give rise to a watery diarrhoea, probably of osmotic origin. Maltose can be absorbed by such patients since maltase 1 and 2 may not be affected, but this does not imply that starch can be tolerated since the lack of isomaltase leads to the accumulation of disaccharides and oligosaccharides with α, $1 \rightarrow 6$ linkages.

In glucose-galactose malabsorption, only the active absorption is affected and fructose is absorbed more rapidly than glucose or galactose. This is in further support of the hypothesis that fructose enters the mucosal cells by a pathway separate from that responsible for the glucose and galactose uptake. Although the sucrase-isomaltase enzyme may be present and fructose absorption proceeds normally, sucrose may not be well tolerated due to the glucose released on hydrolysis.

Malabsorption could theoretically occur in young animals or humans if the enzymes developed at different times. In man, the sucrase-isomaltase enzyme develops during the third month of gestation and is effective at birth. However, man, as a species, is an exception and may be the only mammal equipped at birth with the necessary disaccharidases necessary to digest the sugars present in artificial baby food.[53] The development of absorptive function and the appearance of disaccharidases in the young of various species has been recently reviewed.[56]

SUMMARY

(1) Intestinal handling of carbohydrates involves their enzymatic splitting into simple sugars which are absorbed across the brush border of the mucosal cells.

Of the simple sugars, glucose and galactose are actively absorbed by a carrier mechanism which utilises the existing Na^+ gradient into the cell and which is therefore indirectly dependent on cellular metabolism. The active uptake is specifically and competitively inhibited by the diphenolic glycoside phlorizin.

Fructose and mannose are not actively absorbed but enter the mucosal cell by a separate pathway not dependent on the Na^+ gradient or cellular metabolism and not inhibited by phlorizin.

(2) High molecular weight carbohydrates of the diet (starches) are hydrolysed by the enzyme amylase present in saliva and pancreatic juice, but the disaccharidases which finally split the maltose and oligosaccharides (resulting from amylase activity) into glucose are located in the 'brush border' of the mucosal cells in close spatial relationship to the mechanism for glucose absorption.

One maltase fraction also has sucrase and isomaltase activity and it is this enzyme that is involved in the hydrolysis of dietary sucrose. Of the products of hydrolysis, glucose is almost completely absorbed but fructose is absorbed less efficiently, in keeping with the different capacities for absorption of the simple sugars. However, maltose and sucrose hydrolysis give kinetic advantages to the absorption of the hydrolysis products over what would be expected if the simple sugars were first released into the intestinal lumen.

(3) The physiological characteristics of intestinal digestion and absorption of sugars and the recently recognised malabsorption syndromes can be interpreted and explained by an integration of a number of discrete enzymatic and membrane transport mechanisms in the 'brush borders' of the mucosal cells.

REFERENCES

1. DAHLQVIST, A., *Biochem. J.,* **78**, 282–288 (1961)
2. DAHLQVIST, A. and BORGSTRÖM, B., *Biochem. J.,* **81**, 411–418 (1961)
3. AURICCHIO, S., DAHLQVIST, A. and SEMENZA, G., *Biochim. biophys. Acta,* **73,** 582–587 (1963)
4. AURICCHIO, S., SEMENZA, G. and RUBINO, A., *Biochim. biophys. Acta,* **96,** 498–507 (1965)
5. MILLER, D. and CRANE, R. K., *Biochim. biophys. Acta,* **52**, 281–293 (1961)
6. EICHHOLZ, A. and CRANE, R. K., *Fedn Proc. Fedn Am. Socs exp. Biol.,* **25**, 656 (1966)
7. CRANE, R. K., In: *Carbohydrate metabolism and its disorders,* Ed. by Dickens, Randle and Whelan, Academic Press, London, **1**, 25-51 (1968)
8. JOHNSON, C. F., *Science, N.Y.,* **155**, 1670–1672 (1967)
9. WISEMAN, G., *Absorption from the intestine,* Ed. by Wiseman, Academic Press, London (1964)
10. CORI, C. F., *J. biol. Chem.,* **66**, 691–715 (1925)
11. BÁRÁNY, E. and SPERBER, E., *Skand. Arch. Physiol.,* **81,** 290–299 (1939)
12. FISHER, R. B. and PARSONS, D. S., *J. Physiol., Lond.,* **110,** 36–46 (1949)
13. FISHER, R. B. and PARSONS, D. S., *J. Physiol., Lond.,* **110,** 281–293 (1949)
14. WILSON, T. H. and WISEMAN, G., *J. Physiol., Lond.,* **123,** 116–125 (1954)
15. FISHER, R. B. and PARSONS, D. S., *J. Physiol., Lond.,* **119,** 210–223 (1953)
16. FISHER, R. B. and PARSONS, D. S., *J. Physiol., Lond.,* **119,** 224–232 (1953)
17. WIDDAS, W. F., *J. Physiol., Lond.,* **125**, 163–180 (1954)
18. ALVARADO, F., *Biochim. biophys. Acta,* **109**, 478–494 (1965)
19. WIDDAS, W. F., *J. Physiol., Lond.,* **118**, 23–39 (1952)
20. PARK, C. R., POST, R. L., KALMAN, C. F., WRIGHT, J. H. JR., JOHNSON, L. H. and MORGAN, H. E., *Ciba Fdn Colloq. Endocr.,* **9**, 240–260 (1956)
21. ROSENBERG, T. and WILBRANDT, W., *J. gen. Physiol.,* **41,** 289–296 (1957)
22. McDOUGAL, D. B. JR., LITTLE, K. D. and CRANE, R. K., *Biochim. biophys. Acta.* **45,** 483–489 (1960)
23. CRANE, R. K., In: *Handbook of the Physiology of the Alimentary Canal,* Ed. by Code, Am. Physiol. Soc., Washington, Vol. III, Sect. 6, 1323–1352 (1968)

24. WILSON, T. H. and LANDAU, B. R., *Am. J. Physiol.*, **198,** 99–102 (1960)
25. CRANE, R. K. and KRANE, S. M., *Biochim. biophys. Acta,* **31,** 397–401 (1959)
26. LANDAU, B. R. and WILSON, T. H., *J. biol. Chem.*, **234,** 749–753 (1959)
27. BAILLIEN, M. and SCHOFFENIELS, E., *Biochim. biophys. Acta,* **53,** 537–548 (1961)
28. BARRY, R. J. C., DICKSTEIN, S., MATTHEWS, J., SMYTH, D. H. and WRIGHT, E. M., *J. Physiol., Lond.*, **171,** 316–338 (1964)
29. SCHULTZ, S. G. and ZALUSKY, R., *J. gen. Physiol.*, **47,** 1043–1059 (1964)
30. CRANE, R. K., FORSTNER, G. and EICHHOLZ, A., *Biochim. biophys. Acta,* **109,** 467–477 (1965)
31. RIKLIS, E. and QUASTEL, J. H., *Can. J. Biochem. Physiol.*, **36,** 347–362 (1958)
32. CZÁKY, T. Z. and ZOLLICOFFER, L., *Am. J. Physiol.*, **198,** 1056–1058 (1960)
33. BOSǍCKOVA, J. and CRANE, R. K., *Biochim. biophys. Acta,* **102,** 423–435 (1965)
34. SCHULTZ, S. G., FUISZ, R. E. and CURRAN, P. F., *J. gen. Physiol.*, **49,** 849–866 (1966)
35. SKOU, J. C., *Physiol. Rev.*, **45,** 596–617 (1965)
36. CAPRARO, V., BIANCHI, A. and LIPPE, C., *Archs Sci. biol.*, **47,** 238 (1963)
37. BIHLER, I., *Biochim. biophys. Acta.*, **183,** 169–181 (1969)
38. DARLINGTON, W. A. and QUASTEL, J. H., *Archs Biochem. biophys.*, **43,** 194–207 (1953)
39. WILSON, T. H. and VINCENT, T. N., *J. biol. Chem.*, **216,** 851–866 (1955)
40. HOLDSWORTH, C. D. and DAWSON, A. M., *Clin. Sci.*, **27,** 371–379 (1964)
41. KIYASU, J. Y. and CHAIKOFF, I. L., *J. biol. Chem.*, **224,** 935–939 (1957)
42. NEWEY, H., SANDFORD, P. A. and SMYTH, D. H., *J. Physiol., Lond.*, **168,** 423–434 (1963)
43. WESER, E. and SLEISENGER, M. H., *Gastroenterology,* **48,** 571–578 (1965)
44. GRYBOSKI, J. D., THAYER, W. R. JR., GABRIELSON, I. W. and SPIRO, H. M., *Gastroenterology,* **45,** 633–637 (1963)
45. CAJORI, F. A., *Am. J. Physiol.*, **104,** 659–668 (1933)
46. CHAIN, E. B., MANSFORD, K. R. L. and POCCHIARI, F., *J. Physiol., Lond.*, **154,** 39–51 (1960)
47. DAHLQVIST, A. and THOMPSON, D. L., *J. Physiol., Lond.*, **167,** 193–209 (1963)
48. CRANE, R. K., *Fedn Proc. Fedn Am. Socs exp. Biol.*, **21,** 891–895 (1962)
49. ROTHSTEIN, A., MEIER, R. C. and SCHARFF, T. G., *Am. J. Physiol.*, **173,** 41–46 (1953)
50. TRIER, J. A., In: *Handbook of the Physiology of the Alimentary Canal,* Ed. by Code, Am. Physiol. Soc., Washington, Vol. III, Sect. 6, 1125–1175 (1968)
51. SEMENZA, G., *Protides biol. Fluids,* **15,** 201–208 (1967)
52. SEMENZA, G., CURTIS, C. H., KOLINSKÁ, J. and MÜLLER, M., *Biochim. biophys. Acta,* **146,** 196–204 (1967)
53. SEMENZA, G., In: *Handbook of the Physiology of the Alimentary Canal,* Ed. by Code, Am. Physiol. Soc., Washington, Vol. V, Sect. 6, 2543–2566 (1968)
54. GRAY, G. M. and INGELFINGER, F. J., *J. clin. Invest.*, **45,** 388–398 (1966)
55. DAHLQVIST, A., LINDQUIST, B. and MEEUWISSE, G., In: *Carbohydrate metabolism and its disorders,* Ed. by Dickens, Randle and Whelan, Academic Press, London, **2,** 199–222 (1968)
56. DEREN, J. J., In: *Handbook of the Physiology of the Alimentary Canal,* Ed. by Code, Am. Physiol. Soc., Washington, Vol. III, Sect. 6, 1099–1123 (1968)

Chapter Fifteen

SOME EFFECTS OF DIETARY SUCROSE

ARNOLD E. BENDER and KHATUN B. DAMJI

Reader in Nutrition *Research Assistant in Nutrition*

Department of Nutrition, Queen Elizabeth College, University of London

Attention was drawn to physiological differences between various carbohydrates nearly a century ago. Frerichs, 1876,[1] Külz, 1876,[2] and Lusk, 1892[3] fed fasted animals of various species on glucose, sucrose, fructose, inulin or lactose and showed differences in the amount of glycogen formed in the liver. There are also several early reports indicating that different carbohydrates differ in their fattening effect on animals.[4, 5, 6] More recently the physiological effects of sucrose have been under intensive examination.

Many of the effects of dietary sucrose that have been reported differ with the strain of animal used, their age, the length of time of treatment and the other components of the diet. Consequently some reports that appear to be contradictory may not relate to the same experimental conditions.

STRAIN OF ANIMALS

The response to sucrose of BHE and Holtzman strains of rats are similar but differ from the response of Wistar rats.[7] The first two strains develop larger livers and kidneys, a greater extent of kidney damage and a higher serum cholesterol when fed sucrose instead of stock diet, while the Wistar rats are not affected. Again, sucrose reduces the life span of BHE rats from 22 months to 17 months but has no effect on the life span of Wistar rats.[8]

LENGTH OF TIME ON DIET

Sucrose, compared with starch, increases the level of serum cholesterol after three weeks but the difference disappears after 17 weeks on the sucrose diet.[9] Similarly, sucrose significantly reduces the growth rate of male rats but the differences disappear after 6 months.[10]

AGE AND SEX

Sucrose alters the fat content of the carcass and liver of adult rats[11] although weanlings do not show this effect.[12]

Sex does not appear to be a factor in rats but men, and women after the menopause but not before, respond to dietary sucrose with an increase in serum triglycerides.[13, 14]

OTHER DIETARY FACTORS

The type of fat in the diet affects the response to sucrose.[15] The increase in glucose-6-phosphatase caused by dietary sucrose is less when the dietary fat is hydrogenated coconut oil than when it is corn oil. Similarly, the increase in fructose diphosphatase activity and liver glycogen content following sucrose occurs only when corn oil is fed and not with hydrogenated coconut oil. Another example is the production of fatty livers in adult rats by feeding sucrose in a low-protein diet but not in a high-protein diet.[16]

FOOD CONSUMPTION GROWTH RATE AND BODY COMPOSITION

Differences in the body composition of animals fed different carbohydrates were reported many years ago.[4] It was stated that while sucrose was equal to starch as a direct fuel it was only three quarters as effective in laying down fat.[17] These differences have been shown to be due to faster combustion of sucrose compared with other carbohydrates as indicated by differences in respiratory quotients observed after feeding the different carbohydrates.[5, 6, 18]

There is conflicting evidence of the relative growth rates on animals fed sucrose or starch but the conflict can, to some extent, be explained by the reduced intake of food that often accompanies a sucrose diet. Even on equalised food intakes, however, there is some conflict in the findings. For example, Feyder[17] pair-fed weanling rats on a diet of 18% casein and 68% carbohydrate as sucrose or glucose and found that the former permitted a greater weight gain principally due to fat.

On the other hand Harper and co-workers[19] found no difference between sucrose and dextrin in a paired feeding experiment. In other experiments in which food intake was not measured, these workers found that sucrose had a growth-depressant effect.

In rats, exercised in revolving cages, sucrose and glucose produce the same body composition except for a higher percentage of calcium on the sucrose diet.[20] When glucose is compared with fructose at the same food intake, glucose allows faster growth with more water and protein in the carcasses. This result is complicated by the finding that glucose-fed rats voluntarily undertook more exercise than did fructose-fed animals.[21] There was no difference in weight gains between animals on the two diets.

A further complication is introduced by other factors in the diet, namely fat and protein. The type and amount of fat in the diet affect the amount of food consumed;[22] it also alters the effect of sucrose on liver enzyme and glycogen levels.[15]

Low protein diets emphasise the growth-depressant effect of sucrose and this becomes less marked as the protein level is increased.[23, 24] When dextrin or corn starch is replaced by sucrose in a 9% casein diet there is an increase in the amount of food eaten and a decrease in food conversion efficiency[16, 25, 26] When the protein level is raised to 18% there is no difference between the carbohydrates. Desmoulin,[27] however, found that food consumption decreased in the following order: *(a)* sucrose + 16% casein, *(b)* sucrose + 12% casein, *(c)* starch + 16% casein, *(d)* starch + 12% casein.

The protein effect can be partly quantified, as an extra 2% protein in the diet can compensate for the sucrose and restore the growth rate.[28] Thus 8% fibrin + sucrose permit the same growth as 6% fibrin + dextrin; 10% fibrin + sucrose produce the same growth rate as 8% fibrin + dextrin. These results are only partly explained by differences in food consumption. On the 8% fibrin diets, the calorie intake is less with sucrose than with dextrin, 34 kcal per rat per day compared with 46 kcal. When the extra protein is added to the sucrose diet the food intake is still less on sucrose, 39 kcal per rat per day, but now the growth rates are the same.

There are also changes in body composition; when the extra 2% protein is added to the sucrose diet the sucrose-fed rats gain the same weight as the dextrin-fed animals but their carcasses contain more water, less fat and the same nitrogen. In a similar way no difference is found between glucose and fructose (in a diet containing 18% casein) in their effect on body weight, but fructose produces less body fat, more water and heavier livers.[23] The opposite effect is found in adult rats.[11] Here the feeding of sucrose and fructose, compared with stock diet or glucose diet, results in more body fat.

LIFE SPAN

There are reports that dietary sucrose shortens the life of rats and, as mentioned earlier, there is a strain difference. In a strain of Wistar rats susceptible to renal disease, a diet in which 15% of the starch calories are replaced by sucrose, reduces the average life span of males from 81 weeks to 69 weeks;[10] females are not significantly affected (a reduction of 87 to 83 weeks). Both groups die from renal disease and the onset of the disease is earlier in male rats fed on sucrose. In a strain of Wistar rats susceptible to respiratory disease sucrose has no effect on life span.[8] On the other hand in BHE rats which are susceptible to kidney disease there is a reduction in life span from 88 weeks to 68 weeks.[8]

EFFECTS ON LIVER

Weight—Varying effects of sucrose on liver weight have been reported, the balance of opinion beng that sucrose produces heavier livers than starch[11, 12, 15, 29] although there are reports that sucrose feeding produces lighter livers.[30] In paired-feeding experiments, fructose compared with glucose, gives rise to livers that are 22% heavier, with the same water and protein content but more fat.

Lipids—Both sucrose and fructose result in the accumulation of fat in the livers of weanling rats,[19, 28, 31] in adult rats[11, 32] and rabbits.[33, 34] Sucrose causes an increase in liver triglycerides in male but not in female rats.[35] The fat increase after sucrose feeding is greater in the livers of BHE strain than in those of the Wistar strain.[8] There are contradictory reports of the effects of sucrose feeding on hepatic cholesterol and phospholipid levels.

Glycogen—Reports of the effects of sucrose on liver glycogen are conflicting. Some workers state there is an increase compared with starch[15, 36] others find no difference.[12] Dietary fructose increases the liver glycogen compared with glucose.[37]

Among other effects reported in liver after feeding sucrose, are the increased production of collagen in mice[38] and decreased content of desoxyribonucleic acid.[39]

Metabolism—There have been many investigations of the effect of dietary sucrose upon the subsequent *in vitro* metabolism of glucose by the liver. Liver from rats fed sucrose, and in some instances adipose tissue, show a reduced ability to oxidise glucose and to incorporate glucose into fat.[12, 37] According to different reports the incorporation of acetate into fat is increased by sucrose feeding,[10] decreased[41] or not changed.[42, 43]

It is possible that, as with other measurements already discussed, these contradictory results are due to the use of different strains of animals. Bender and Damji[44] found that, out of five strains of rats fed sucrose, three responded by a depressed liver metabolism *in vitro* while two strains did not respond.

KIDNEYS

In some strains of rats the kidneys are affected by dietary sucrose. In the BHE strain in which renal disease is common, sucrose causes earlier death and the deaths are due to renal disease.[8] The Wistar strain of rats, which normally die of respiratory disease, have only a slight increase in kidney size compared with the BHE strain.

Rats fed sucrose show enlarged kidneys and injection with radioactive glucose showed reduced fat synthesis in the kidneys.[12]

SERUM LIPIDS

Animals—Several workers have shown that when sucrose is fed together with cholesterol there is an increase in serum cholesterol in rats,[45] chickens[46, 47] and rabbits[48] although the changes in rats are less striking than those in other species.

The elevated serum cholesterol found in chicks does not occur in germ-free animals.[47] On the other hand, glucose causes a 40% increase in serum cholesterol in germ-free animals. When antibiotics are administered to conventional chicks the results are similar to those obtained with germ-free animals.

In order to obtain an increase in cholesterol levels it appears to be necessary to feed cholesterol together with the sucrose.[49, 50]

Man—There are many reports of the effects of dietary sucrose on serum lipids in man,[51, 52] both on cholesterol and triglyceride levels. The effect on cholesterol, however, is not specifically due to sucrose as other simple carbohydrates, such as glucose and lactose have the same effect.[53, 54] The level of fat influences the effect of sucrose, which is more marked when fat intake is low and minimal when fat intake is high.[55–60]

The level of triglycerides appears to be increased by sucrose more consistently than does that of cholesterol; sucrose but not starch, maltose, glucose or liquid glucose causes a rise in serum triglycerides in men[61, 62, 63] An increase in the amount of dietary sucrose caused an increase in triglycerides but no increase in cholesterol or phospholipids.[64, 65]

INSULIN AND BLOOD SUGAR

It has been claimed that the feeding of sucrose for a period of 6–8 weeks causes an impairment of glucose tolerance in rats[66] and also in human subjects.[67, 68] These findings appear to be linked with insulin. First the reduction in blood glucose in rats following an injection of tolbutamide is less in animals fed sucrose than in those fed starch.[69] Secondly, an increase in serum immuno-reactive insulin occurred in 6 out of 19 human subjects fed sucrose; the same subjects showed an increase in blood platelet adhesiveness and a weight increase.[65] In patients with peripheral vascular disease, serum insulin levels and platelet adhesiveness correlate with sucrose intake whereas such a correlation is not seen with normal subjects.[70]

OTHER EFFECTS

It was shown many years ago[5] that the respiratory quotient (*RQ*) rose more rapidly after the administration of fructose than after glucose. As the *RQ* after fructose was frequently greater than unity, these authors concluded that fructose is more easily converted into fat than is glucose. Higgins[6] found that when 100 g dose of fructose or sucrose was given to a human subject there was a rise in *RQ* in 4–8 min while glucose and maltose caused a rise only after 20–30 min. Fructose but not glucose causes a rise in blood lactate, although this does not account for the difference in *RQ* between the sugars.[71]

Evidence from a number of workers suggests that sucrose stimulates carbohydrate degradation since a dose of fructose produces a greater rise in blood and urine lactic acid[72, 73] and also in blood pyruvate than does glucose.[74]

SUGGESTED EXPLANATIONS

Since some of the effects of sucrose described above appear to be very different in nature it would seem that more than one mechanism of action is involved. Several explanations have been put forward of which three will be discussed briefly, namely, *(a)* that the sucrose effect compared with starch is due to the different metabolic routes of fructose and glucose, *(b)* that the different carbohydrates are absorbed at different rates and *(c)* that sucrose passes along the intestine more rapidly than does starch.

(a) If the effect of sucrose is due to the fructose moiety then the effects of dietary sucrose and fructose should be similar and different from those of starch and glucose. The effects of sucrose and of fructose on the levels of many enzymes are indeed the same, and the finding that glucose has little or no effect on these enzyme levels indicates that fructose follows a different metabolic pathway from that of glucose. For example glucose-6-phosphatase, glucose-6- phosphate dehydrogenase, phosphogluconate dehydrogenase, aldolase, fructokinase, malic dehydrogenase and α-glycerophosphate dehydrogenase are increased by dietary fructose but not glucose.[75–79] On the other hand glucokinase is elevated more by glucose than by fructose.[80]

These changes in enzyme levels could be responsible for the changes in fat metabolism and glycogen storage that have been shown to accompany dietary sucrose. Thus Tepperman and Tepperman[81, 82] suggested that the increased levels of glucose-6-phosphate dehydrogenase and phosphogluconate dehydrogenase indicate increased utilisation of the hexose monophosphate shunt and that this would increase the amount of reduced NADP formed and consequently stimulate fat synthesis.

Pyruvate kinase is considerably increased by sucrose feeding[83, 84] and this enzyme plays a key part in the switch-over from glycogen storage to fat synthesis. Sucrose and fructose, but not glucose, increase hepatic glucose-6-phosphatase and so would reduce the pool of glucose-6-phosphate and could account for the *in vitro* depression of glucose oxidation caused by feeding sucrose.[12, 75]

(b) Rate of absorption. If the difference in metabolism of fructose and glucose is to be offered as the explanation of the different metabolic effects of dietary sucrose and starch, then dietary fructose would cause the same effects as sucrose. As stated above this is true when measured by many parameters but in some instances it is not so. For example after oral doses of radioactive sugars, the specific activity of the carbon dioxide from sucrose and maltotriose was three times as great as that from glucose, fructose or a mixture of the two.[85]

It is suggested that differences of this kind, i.e. between sucrose on the one hand and glucose and fructose on the other are due to the different rates of absorption.[86, 87] Certainly there is evidence that the type of carbohydrate influences the emptying time of the stomach.[88, 89] The rate of entry of the different sugars into the blood stream would have an effect on insulin secretion, which has already been suggested as a major factor.

(c) Rate of passage along the intestine. In the early days of vitamin and amino acid studies, it was noted that when starch provided the carbohydrate of the ration it was more difficult to produce B vitamin deficiency symptoms than when sucrose was used.[24, 90, 91, 92] This effect of sucrose appears to be quite different from the enzyme-linked effects on fat and carbohydrate metabolism discussed above and could be readily explained by the slower passage of starch along the intestine. This would promote the growth of bacteria in the large intestine and the synthesis of the B vitamins. In fact Guerrant and Dutcher[90] demonstrated that there were less of the B vitamins in the faeces of sucrose-fed rats compared with starch-fed animals. As rats practise coprophagy, the vitamins would be available to the animal. There is no evidence that sucrose and starch have different effects on the vitamin requirements of man.

Amino acid metabolism is also affected by the dietary carbohydrate. For example, when the carbohydrate of the diet is changed from sucrose to starch, the nitrogen balance on low levels of dietary protein changes from negative to equilibrium[93] and extra dietary protein can compensate for the growth-depressant effect of sucrose.[26, 28] It is suggested that the slower passage of starch along the intestine, with the consequently more continuous absorption compared with that of sucrose, may provide a protein-sparing effect.[26, 28]

REFERENCES

1. FRERICHS, E., *Dissert. Würzburg,* Fortschritte der Thier-Chemie, 198 (1876)
2. KÜLZ, E., *Sber. Ges. Beförd. ges. Naturw. Marburg,* **5,** 201 (1876)
3. LUSK, G., *Z. Biol.,* **28,** 245 (1892)
4. KELLNER, O., 1909. *Die Verfütterung der zuckerfuttermittel,* Arbeiten d. Deutschen Landw-Gessel. Berlin. Hft. 152 (Cited by Wendt in 1923) (1909)
5. TÖGEL, O., BREZINA, E. and DURIG, A., *Biochem. Z.,* **1,** 296 (1913)
6. HIGGINS, H. L., *Am. J. Physiol.,* **41,** 258 (1916)
7. MARSHALL, M. W. and HILDEBRAND, H. E., *J. Nutr.,* **79,** 227 (1963)
8. DURAND, A. M., FISHER, M. and ADAMS, M., *Archs Path.,* **85,** 318 (1968)
9. FILLIOS, L. C., NAITO, C., ANDRUS, S. B., POSTWAN, O. W. and MARTIN, R. S., *Am. J. Physiol.,* **194,** 275 (1958)
10. DALDERUP, L. M. and VISSER, W., *Nature, Lond.,* **222,** 1050 (1969)
11. ALLEN, R. J. L. and LEAHY, J. S., *Br. J. Nutr.,* **20,** 339 (1966)
12. BENDER, A. E. and THADANI, P. V., *Nutritio Dieta,* **12,** 22 (1970)
13. MACDONALD, I., *J. clin. Nutr.,* **18,** 86 (1966)
14. COLTART, T. M., *Nature, Lond.,* **222,** 575 (1969)
15. CARROLL, C., *J. Nutr.,* **79,** 93 (1963)
16. AL NAGDY, S., MILLER, D. S., QURESHI, R. U. and YUDKIN, J., *Nature, Lond.,* **209,** 81 (1966)
17. FEYDER, S., *J. Nutr.,* **9,** 457 (1935)
18. EDWARDS, H. T., BENSLEY, E. H., DILL, D. B. and CARPENTER, T. M., *J. Nutr.,* **27,** 241 (1944)
19. HARPER, A. E., MONSON, W. J., ARATA, D. A., BENTON, D. A. and ELVEHJEM, C. A., *J. Nutr.,* **51,** 523 (1953)
20. MITCHELL, H. H., HAMILTON, T. S. and BEADLES, J. R., *J. Nutr.,* **14,** 435 (1937)
21. HALDI, J., BACHMANN, G., ENSOR, C. and WYNN, W., *J. Nutr.,* **16,** 239 (1938)
22. MARSHALL, M. W., HILDEBRAND, H. E., DUPONT, J. I. and WOMACK, M., *J. Nutr.,* **69,** 371 (1959)
23. BACHMANN, G., HALDI, J., WYNN, W. and ENSOR, C., *J. Nutr.,* **16,** 229 (1938)
24. KREHL, W. A., SARMA, P. S., TEPLY, L. J. and ELVEHJEM, C. A., *J. Nutr.,* **31,** 85 (1946)
25. HARPER, A. E., MONSON, W. J., BENTON, D. A. and ELVEHJEM, C. A., *J. Nutr.,* **50,** 383 (1953)
26. HARPER, A. E. and KATAYAMA, M. C., *J. Nutr.,* **49,** 261 (1953)
27. DESMOULIN, B., *Annls. Biol. anim. Biochim. Biophys.,* **7,** 163 (1967)
28. WEINER, R. P., YOSHIDA, M. and HARPER, A. E., *J. Nutr.,* **80,** 279 (1963)
29. ADAMS, M., FISHER, M. and KOVAL, J. G., *Fedn Proc. Fedn Am. Socs exp. Biol.,* **18,** 178 (1959)
30. GUGGENHEIM, K., ILAN, J., PERETZ, E., *J. Nutr.,* **72,** 93 (1960)

31. LITWACK, G., HANKES, L. V. and ELVEHJEM, C. A., *Proc. Soc. exp. Biol. Med.*, **81,** 441 (1952)
32. MARSHALL, M. W. and WORMACK, M., *J. Nutr.*, **57,** 193 (1955)
33. MACDONALD, I., *J. Physiol., Lond.*, **160,** 306–316 (1962)
34. MACDONALD, I., *J. Physiol., Lond.*, **162,** 334 (1962)
35. MORTON, R. A. and HORNER, A. A., *Biochem. J.*, **79,** 631 (1961)
36. GAERTNER, E. B. and CARROLL, C., *J. Nutr.*, **91,** 69 (1967)
37. HILL, R., BAKER, N. and CHAIKOFF, I. L., *J. biol. Chem.*, **209,** 705 (1954)
38. ZUCKERMAN, A. J. and MACDONALD, I., *Br. J. exp. Path.*, **45,** 589 (1964)
39. BEDO, M. and SZIGETI, A., *Nährung,* **11,** 305 (1967)
40. FABRY, P., POLEDNE, R., KAZDOVA, L. and BRAUN, T., *Nutritio Dieta,* **10,** 81 (1968)
41. BAILEY, E., TAYLOR, C. B. and BARTLEY, W., *Nature, Lond.*, **217,** 472 (1968)
42. ZAKIM, D. and HERMAN, R. M., *Am. J. clin. Nutr.*, **21,** 315 (1968)
43. BAR-ON, H. and STEIN, Y., *J. Nutr.*, **94,** 95 (1968)
44. BENDER, A. E. and DAMJI, K. B. *(unpublished)*
45. PORTMAN, O. W., LOWRY, E. Y. and BRUNO, D., *Proc. Soc. exp. Biol. Med.*, **91,** 321 (1956)
46. GRANT, W. C. and FAHRENBACH, M. J., *Fedn Proc. Fedn Am. Socs exp. Biol.*, **16,** 50 (1957)
47. KRITCHEVSKY, D., GRANT, W. C., FAHRENBACH, M. J., RICCARDI, B. A. and McCANDLESS, R. F. J., *Archs Biochem. Biophys.*, **75,** 142 (1958)
48. GRANT, W. C. and FAHRENBACH, M. J., *Proc. Soc. exp. Biol. Med.*, **100,** 250 (1959)
49. STAUB, H. W. and THIESSEN, R., *J. Nutr.*, **95,** 633 (1968)
50. POLLOCK, O. J., *Circulation,* **18,** 488 (1958)
51. KEYS, A., ANDERSON, J. T. and GRANDE, F., *J. Nutr.*, **70,** 257 (1960)
52. LOPEZ, A., HODGES, R. E. and KREHL, W. A., *Am. J. clin. Nutr.*, **18,** 149 (1966)
53. ANDERSON, J. T., GRANDE, F., MATSUMOTO, Y. and KEYS, A., *J. Nutr.*, **79,** 349 (1963)
54. SHAMMA'A, M. and AL-KHALIDI, U., *Am. J. clin. Nutr.*, **13,** 194 (1963)
55. HEGSTED, D. M., McGANDY, R. B., MYERS, M. L. and STARE, F. J., *Am. J. clin. Nutr.*, **17,** 281 (1965)
56. STARE, F. J., *J. Am. diet. Ass.*, **48,** 8 (1966)
57. McGANDY, R. B. HEGSTED, D. M., MYERS, M. L. and STARE, F. J., *Am. J. clin. Nutr.*, **18,** 237 (1966)
58. KUO, P. T. and BASSETT, D. R., *Chem Abstr.*, **63,** 3437g (1965)
59. GROEN, J. J., COHEN, A. M., BALOGH, M. and YARON, E., *Am. J. clin. Nutr.*, **19,** 46 (1966)
60. KAUFMANN, N. A., POZNANSKI, R., BLONDHEIM, S. H. and STEIN, Y., *Am. J. clin. Nutr.*, **18,** 261 (1966)
61. MACDONALD, I. and BRAITHWAITE, D. M., *Clin. Sci.*, **27,** 23 (1964)
62. MACDONALD, I., *Clin. Sci.*, **29,** 193 (1965)
63. MACDONALD, I., *Am. J. clin. Nutr.*, **18,** 369 (1966)
64. AKINYANJU, P. A., *Ph.D. Thesis*, Faculty of Science, University of London (1968)
65. SZANTO, S. and YUDKIN, J., *Postgrad. med. J.*, **45,** 602 (1969)
66. COHEN, A. M. and TEITELBAUM, A., *Am. J. Physiol.*, **206,** 105 (1964)
67. COHEN, A. M., TEITELBAUM, A., BALOGH, M. and GROEN, J. J., *Am. J. clin. Nutr.*, **19,** 59 (1966)
68. COHEN, A. M., *Am. J. clin. Nutr.*, **20,** 126 (1967)
69. COHEN, A. M. and YUDKIN, J., *Biochim. biophys. Acta*, **141,** 637 (1967)
70. SZANTO, S., YUDKIN, J. and KAKKAR, V. V., *Postgrad. med. J.*, **45,** 608 (1969)
71. BACHMANN, G. and HALDI, J., *J. Nutr.*, **13,** 157 (1937)

72. CAMPBELL, W. R. and MALTBY, E. T., *J. clin. Invest.*, **6**, 303 (1928)
73. WIERZUCHOWSKI, M. and LANIEWSKI, M., *Biochem. Z.*, **230**, 173 (1931); *Nutr. Abstr. Rev.*, **1**, 159 (1931)
74. DODDS, C., MILLER, A. L. and ROSE, C. F. M., *Lancet*, **2**, 178 (1960)
75. FREEDLAND, R. A. and HARPER, A. E., *J. biol. Chem.*, **230**, 833 (1958)
76. FITCH, W. M., HILL, R. and CHAIKOFF, I. L., *J. biol. Chem.*, **234**, 1048 (1959)
77. FITCH, W. M. and CHAIKOFF, I. L., *J. biol. Chem.*, **235**, 554 (1960)
78. SZEPASI, B. and FREEDLAND, R. A., *J. Nutr.*, **94**, 37 (1968)
79. AITKEN, J., *M.Sc. Thesis*, Faculty of Science, University of London (1969)
80. BLUMENTHAL, M. D., ABRAHAM, S. and CHAIKOFF, I. L., *Archs Biochem. Biophys.*, **104**, 215 (1964)
81. TEPPERMAN, J. and TEPPERMAN, H. M., *Am. J. Physiol.*, **193**, 55 (1958)
82. TEPPERMAN, J. and TEPPERMAN, H. M., *Diabetes*, **7**, 478 (1958)
83. KREBS, H. A. and EGGLESTON, L. V., *Biochem. J.*, **94**, 3c (1965)
84. YUDKIN, J. and KRAUSS, R., *Nature, Lond.*, **215**, 75 (1967)
85. MARTIN, A. F. M. and YOUNG, F. G., *Nature, Lond.*, **215**, 885 (1967)
86. MACDONALD, I. and TURNER, L. J., *Lancet*, **i**, 841 (1968)
87. NAISMITH, D. J., *unpublished* (1970)
88. PERAINE, C., ROGERS, Q. R., YOSHIDA, M., CHEN, M. and HARPER, A., *Can. J. Biochem. Physiol.*, **37**, 1475 (1959)
89. MEHNERT, H., and FORSTER, H., *Diabetologia*, **4**, 26 (1968)
90. GUERRANT, N. B. and DUTCHER, R. A., *J. Nutr.*, **8**, 397 (1934)
91. MANNERING, G. J., ORSINI, D. and ELVEHJEM, C. A., *J. Nutr.*, **28**, 141 (1944)
92. HUNDLEY, J. M., *J. biol. Chem.*, **181**, 1 (1949)
93. WOMACK, M., MARSHALL, M. W. and PARKS, A. B., *J. Nutr.*, **51**, 117 (1953)

Chapter Sixteen

THE HYPERLIPIDAEMIC PROPERTY OF SUCROSE

DONALD J. NAISMITH
Lecturer in Nutrition, Queen Elizabeth College, University of London

Atherosclerosis, a disease characterised by the accumulation of lipids in the subendothelial space of the larger blood vessels, cannot usually be detected in living people. The disease progresses silently over many years, and is recognised only when one of its clinical complications, such as myocardial infarction, is established, and this may be too late to be of much benefit to the patient.

When it was first reported by Keys and his associates[1] that patients with coronary heart disease tend to have relatively high concentrations of cholesterol in the blood, and that this peculiarity precedes the appearance of the clinical signs of heart disease, a simple biochemical test, the estimation of serum cholesterol, became available for assessing whether a person was likely to be suffering from atherosclerosis or not. While Keys' prospective study was in progress, Albrink and her associates[2] discovered that patients who have had a myocardial infarction also have raised serum triglyceride concentrations. And so the measurement of the degree of lipidaemia, the blood lipid concentrations, came into use for diagnosing atherosclerosis and for predicting the risk of future coronary heart disease, although a direct cause and effect relationship between the rise in blood lipids and the degenerative changes which occur in the artery walls has yet to be proved.

As a corollary to this relationship between blood lipids and the incidence of atherosclerosis, it was argued that any factor, dietary or otherwise, which was known to influence blood lipid values could be regarded either as a harmful agent or as a prophylactic agent, depending on whether it promoted a rise or fall in the blood lipid concentrations.

The case against sucrose has been presented by Professor Yudkin *(see Chapter 20)*. It has been repeatedly shown, in our department and in other laboratories,[3–8] that when part or all of the starch in the diet of experimental animals or of man is replaced with sugar, a rise occurs in the plasma concentration of triglycerides and, under certain circumstances, of cholesterol.

An explanation for this hyperlipidaemic property of sucrose has been the aim of research carried out by me and my colleague, Dr. I. Rana. Two facts that had been established in other laboratories provided the starting point for our investigation. Firstly, plasma triglycerides originate in the liver,[9–12] the fatty acids being derived either from dietary fat, or from synthesis from carbohydrates. Secondly, fructose, a constituent monosaccharide of sucrose, but not of starch, is metabolised to a very limited degree outside the liver.[13, 14]

I therefore set up the hypothesis that the difference in the plasma lipid response to the feeding of sucrose and starch is a consequence of a difference in the site of metabolism of these carbohydrates; that monosaccharides entering the blood following the ingestion of starch are largely metabolised in the extra-hepatic tissues, whereas the monosaccharides derived from sucrose are preferentially metabolised in the liver. Thus, with starch, fat is synthesised primarily in adipose tissue, while with sucrose, fat synthesis proceeds in the liver, and triglycerides are thereafter mobilised from the liver to the adipose tissue. The elevation in the plasma triglyceride concentrations in the sucrose-fed animal is a concomitant of this augmented through-put of triglycerides in the plasma.

In the first experiments designed to test this hypothesis, rats were fed, for 50 days, diets containing either sucrose or starch as the major energy source. Two other groups were given diets in which most of the non-protein calories were derived from an animal or vegetable fat. Our aim was to relate the fasting lipid concentrations in the plasma at the end of the feeding period to the rate of lipid (fatty acid) synthesis in various tissues.

As an index of lipogenesis we measured the activities of two enzymes, pyruvate kinase and glucose-6-phosphate dehydrogenase. The idea that changes in enzyme activity reflect, quantitatively, changes in throughput in the metabolic pathway in which the enzyme participates was first put forward by Yudkin.[15] The work of Fitch and Chaikoff[16] on adaptation to high carbohydrate diets, in which alterations in metabolic pathways postulated on the basis of enzyme measurements were verified by isotopic techniques, contributed greatly to the acceptance of such measurements as a valid experimental approach.

Sucrose feeding has been shown to increase the activities of a number of key hepatic enzymes of the glycolytic and hexose monophosphate oxidative pathways of glucose metabolism, and of lipogenesis in the liver. The changes induced by sucrose fall into three categories: those specifically associated with the metabolism of fructose and the conversion of fructose to glucose,[17–20] those occurring in response to the feeding of a high carbohydrate diet[21, 22] and those resulting from a change in the nature of the dietary carbohydrate, namely the substitution of sucrose for starch.[20, 23]

The enzymes pyruvate kinase and glucose-6-phosphate dehydrogenase are involved in the biosynthesis of fatty acids at different stages. The former determines the rate of production of pyruvate, and hence acetyl CoA, the basic building block from which fatty acids are constructed, while the latter regulates the production of reducing power, in the form of NADPH, which is consumed in the condensation of acetyl CoA units to form long chain fatty acids.

The fasting plasma lipid concentrations of rats fed the high carbohydrate or high fat diets are shown in *Table 1*, and in

Table 1

Diet	*Triglycerides* (mg/100 ml)	*Total cholesterol* (mg/100 ml)
Starch	42·5	157
Sucrose	60·1*	174*
Sunflower seed oil	27·0*	165
Dripping	31·6*	157

*Value significantly different from that for starch.

Table 2 are shown the activities of pyruvate kinase (PK) and glucose-6-phosphate dehydrogenase (G-6-PD) in the tissues. The values are the means for 10 litter-matched animals in each experimental group. The starch-fed group acts as a control.

Feeding sucrose caused a rise in the plasma triglycerides and also in cholesterol. Both of the fats reduced the plasma triglyceride concentration, but did not affect the cholesterol level.

Table 2

Diet	*Liver* (I.U./*liver*)		*Adipose tissue* (I.U./g)		*Muscle* (I.U./g)
	PK	G-6-PD	PK	G-6-PD	PK
Starch	154	23·6	2·0	0·33	88
Sucrose	312*	37·8*	1·8	0·24*	94
Sunflower seed oil	39*	3·3*	1·0*	0·11*	98
Dripping	56*	4·3*	0·7*	0·10*	96

*Value significantly different from that for starch.

In the liver, sucrose feeding doubled the activities of the two enzymes. The dietary fats, on the other hand, greatly reduced enzyme activities. This was not unexpected; when the diet provides an ample source of fatty acids, endogenous fat synthesis is suppressed.

In the adipose tissue, sucrose depressed G-6-PD activity, and appeared to depress PK activity, but the difference in values between the sucrose and starch-fed groups did not in this case reach statistical significance. In this tissue also, high-fat diets reduced endogenous fat synthesis. Muscle PK did not seem to participate in these metabolic adaptations.

These experiments clearly demonstrated that a rise in blood triglycerides is associated with increased hepatic lipogenesis, and, conversely, a fall in the plasma concentration accompanies depressed hepatic lipogenesis. Furthermore, as postulated, fatty acid synthesis in the adipose tissue is reduced when sucrose is fed.

The problem was then to determine why the replacement of dietary starch with sucrose should lead to an increase in lipid synthesis in the liver. Two explanations were considered. The fructose component of the sucrose molecule might be more readily converted to fat than is glucose derived from starch. Alternatively, if sucrose were more rapidly digested and absorbed, monosaccharides entering the liver at a higher concentration might lead to adaptive changes in the activities of the enzymes involved in their metabolism.

These ideas were examined in the next experiments, in which a variety of carbohydrates were fed; starch, maltose and glucose, all of which supply only glucose to the tissues, and sucrose and fructose.

After 50 days on diet, the animals were killed, the blood was analysed for lipids, and the activities of pyruvate kinase and glucose-6-phosphate dehydrogenase in the livers and adipose tissue were measured. The results of the analyses are summarised in *Tables 3* and *4*.

Table 3

Diet	*Triglycerides* (mg/100 ml)	*Total cholesterol* (mg/100 ml)
Starch	51	111
Glucose	77*	118
Maltose	81*	128*
Sucrose	88*	124*
Fructose	78*	135*

*Value significantly different from that for starch.

All of the sugars promoted an increase in the plasma triglyceride concentration of roughly the same order, and all but glucose significantly raised the cholesterol value.

Table 4

	Liver (I.U./*liver*)		*Adipose tissue* (I.U./g)	
	PK	G-6-PD	PK	G-6-PD
Starch	93	20	1·56	0·31
Glucose	148*	29*	1·66	0·29
Maltose	196*	37*	2·02	0·37
Sucrose	198*	37*	1·14	0·26
Fructose	376*	60*	0·65*	0·11*

*Value significantly different from that for starch.

Likewise, all the sugars raised the activities of both enzymes in the liver. Maltose proved to be a more potent stimulus to hepatic lipogenesis than was glucose, giving values very similar to those for sucrose, while fructose gave the highest value of all.

In the adipose tissue, when fructose was the sole dietary carbohydrate, fatty acid synthesis was markedly depressed.

When sucrose was fed, the values for the activities of both enzymes were again lower than those found for starch, but differences attained statistical significance only when the number of littermate pairs of rats was raised to fifteen. It should be borne in mind that one half of the dietary sucrose, following hydrolysis in the brush border of the intestinal mucosa, is immediately made available to the tissues as glucose, and that a further amount of glucose is derived from fructose in the intestinal mucosa[24] and in the liver.[25] The adipose tissue is thus deprived of glucose to a much lesser extent when sucrose, rather than fructose, is used as the dietary source of non-protein calories.

The results of these experiments suggest that the primary stimulus to enhanced hepatic lipogenesis is a recurrent flooding of the liver cells with high concentrations of monosaccharides, which leads to the induction of enzymes involved in fatty acid synthesis, and that any differences in the pathways of fructose and glucose metabolism are of minor importance in determining the hyperlipidaemic effect of sucrose.

This opinion was confirmed in a final experiment in which the same carbohydrates as previously used were again fed to rats for 50 days. A single 2 g meal of the diets to which the animals had been accustomed was then fed, and blood was taken from the tail at 5 min intervals thereafter, for estimations of total reducing sugars and fructose. The 'meal tolerance' curves are shown in *Figure 1*.

Differences in the shapes of these curves are accounted for by the interplay of three factors:

(i) The rate of entry of monosaccharides into the portal vein, and release from the liver into the hepatic vein,

(ii) the proportion of the monosaccharides diverted in the liver for fatty acid synthesis, and

(iii) the rate of uptake of the monosaccharides from systemic blood by the extra-hepatic tissues.

The rise in blood sugar level was much more rapid following the ingestion of meals containing sucrose, maltose and glucose than it was after a starch meal.

The lower concentration of sugars after a fructose meal is attributed to the greater proportion of this sugar retained by the liver for lipogenesis, as evidenced by the threefold and fourfold increase in the activities of hepatic glucose-6-phosphate dehydrogenase and pyruvate kinase.

Maltose appeared to undergo more rapid digestion and absorption than did glucose, and the more rapid fall in blood glucose concentration following maltose ingestion confirms that more of the glucose derived from maltose was being converted into fat in the liver.

Finally the failure of the blood sugar concentration to fall some 15 min after a fructose or sucrose containing meal, as

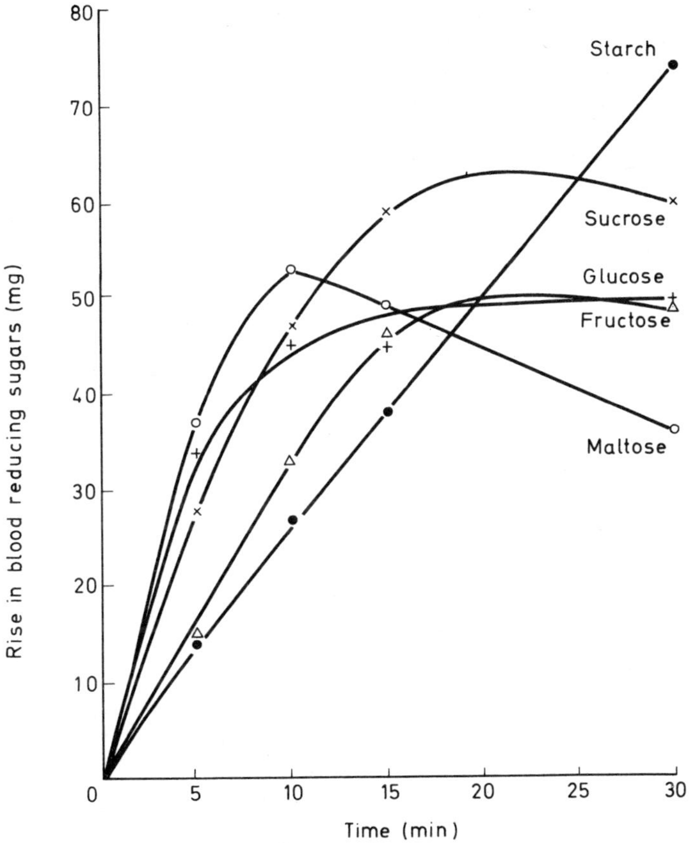

Figure 1. Rise in blood reducing sugars following meals containing different carbohydrates

it did in the case of maltose-fed rats, is due to the fact that a considerable part of the reducing sugar was in the form of fructose, which, as mentioned earlier, is not removed from circulation by the extra-hepatic tissues.

The few sucrose or fructose tolerance tests that have been conducted on human volunteers[26, 27] have shown that neither of these sugars, given by mouth, produces as great a rise in blood reducing sugars as does glucose or 'liquid' glucose (partially hydrolysed starch); the increase in hexose sugars in the blood following fructose administration was less than half that found with glucose or 'liquid' glucose, and approximately half of this increase was due to fructose. Furthermore a significant rise in blood pyruvate was observed with both sucrose and fructose, which may be taken as an indication of augmented through-put of sugars in the Embden-Meyerhoff pathway in the liver.

It appears, therefore, that in man, as in the rat, a substantial part of ingested sucrose does not reach the systemic blood as monosaccharides, but is diverted by the liver, presumably for fat synthesis, and that hyperlipidaemia encountered in human volunteers consuming sucrose-rich diets may also be explained on this basis.

REFERENCES

1. KEYS, A., TAYLOR, H. L., BLACKBURN, H., BROZEK, J., ANDERSON, J. T. and SIMONSON, E., *Circulation,* **28,** 381 (1963)
2. ALBRINK, M. J., MEIGS, J. W. and MAN, E. B., *Am. J. Med.,* **31,** 4 (1961)
3. AKINYANJU, P. A. and YUDKIN, J., *Proc. Nutr. Soc.,* **26,** xxxi (1967)
4. SZANTO, S. and YUDKIN, J., *Postgrad. med. J.,* **45,** 602 (1969)
5. MACDONALD, I. and BRAITHWAITE, D. M., *Clin. Sci.,* **27,** 23 (1964)
6. MACDONALD, I., *Am. J. clin. Nutr.,* **16,** 458 (1965)
7. ALLEN, R. J. L. and LEAHY, J. S., *Br. J. Nutr.,* **20,** 339 (1966)
8. TAYLOR, D. D., CONWAY, E. S., SCHUSTER, E. M. and ADAMS, M., *J. Nutr.,* **91,** 275 (1967)
9. HARPER, P. V., JR., NEAL, W. B., JR. and HLAVACEK, G. R., *Metabolism,* **2,** 69 (1953)
10. BYERS, S. O. and FRIEDMAN, M., *Am. J. Physiol.,* **198,** 629 (1960)
11. BORGSTRÖM, B. and OLIVECRONA, T., *J. Lipid Res.,* **2,** 263 (1961)
12. HAVEL, R. J. and GOLDFIEN, A., *J. Lipid Res.,* **2,** 389 (1961)
13. WICK, A. N., SHERRILL, J. W. and DRURY, D. R., *Diabetes,* **2,** 465 (1953)
14. HERMAN, R. H. and ZAKIM, D., *Am. J. clin. Nutr.,* **21,** 245 (1968)
15. YUDKIN, J., *L. Farkas Memorial Volume, Special Publication No. 1,* Research Council of Israel, Jerusalem (1952)
16. FITCH, W. M. and CHAIKOFF, I. L., *J. biol. Chem.,* **235,** 554 (1960)
17. CARROLL, C., *J. Nutr.,* **79,** 93 (1963)
18. AITKEN, J., ROBINSON, D. and YUDKIN, J., *Proc. Nutr. Soc.,* **26,** xxx (1967)
19. STIFEL, F. B., ROSENSWEIG, N. S., ZAKIM, D. and HERMAN, R. H., *Biochim. biophys. Acta,* **170,** 221 (1968)
20. BARTLEY, W., DEAN, B., TAYLOR, C. B. and BAILEY, E., *Biochem. J.,* **103,** 550 (1967)
21. KREBS, H. A. and EGGLESTON, L. V., *Biochem. J.,* **94,** 3C (1965)
22. KONISHI, F., *J. Nutr.,* **89,** 329 (1966)
23. YUDKIN, J. and KRAUSS, R., *Nature, Lond.,* **215,** 75 (1967)

24. KIYASU, J. Y. and CHAIKOFF, I. L., *J. biol. Chem.*, **224,** 935 (1957)
25. MILLER, L. L., BURKE, W. T. and HAFT, D. E., *Fedn Proc. Fedn Am. Socs exp. Biol.*, **14,** 707 (1955)
26. BUTTERFIELD, W. J. H., SARGEANT, B. M. and WHICHELOW, M. J., *Lancet,* **1,** 574 (1964)
27. SWANN, D. C., DAVIDSON, P. and ALBRINK, M. J., *Lancet,* **1,** 60 (1966)

Chapter Seventeen

SUCROSE AND BLOOD LIPIDS

IAN MACDONALD
Professor of Applied Physiology, Guy's Hospital Medical School, London

The comparatively recent finding that an increase in the concentration of triglyceride in the fasting serum is more likely to be induced by dietary carbohydrate than by dietary fat[1] has turned the spotlight on the role of dietary carbohydrates in those diseases associated with a raised level of fasting serum triclyceride. Such diseases include atherosclerosis[2] and diabetes mellitus.[3]

The part played by dietary carbohydrates in determining the level of cholesterol and phospholipid in the fasting serum is not so clear, but it is likely that an elevation in serum triglyceride concentration is also accompanied by an elevation in cholesterol and phospholipid concentration since the lipo-protein fraction that accompanies endogenous triglyceride also carries with it smaller proportions of cholesterol and phospholipid. It is feasible that certain dietary carbohydrates may, in some way, influence the level of cholesterol to a greater extent than would be expected by the increase in the lipo-protein fraction associated with triglycerides.

The physiological chemist, until recently, considered that the metabolic effects of the various types of dietary carbohydrate consumed were identical. It was considered that fructose and galactose, the common monosaccharides apart from glucose, were converted to glucose before being metabolised. These views were held despite isolated evidence to the contrary. For example, in 1913 it was reported[4] that the respiratory quotient following the ingestion of fructose tended to be greater than 1·0 whereas after glucose the level was persistently less than 1·0. Six years later, Higgins,[5] as a result of experiments in man in which he also measured the respiratory quotient, concluded that there was 'a fundamental and distinct difference in the metabolism of the various sugars in man'. He stated that fructose shows a tendency or preference to change to fat in the body while glucose tends to change to glycogen and be stored as such.

Evidence has been slowly accumulating that the lipid metabolism of the organism is not affected in the same way by all dietary carbohydrates, and the recent description of a carbohydrate-induced hypertriglyceridemia has raised the question as to whether all carbohydrates are equal in this respect.

Sucrose is a very commonly consumed carbohydrate and its world wide consumption seems to be increasing though perhaps its *per capita* consumption in some countries such as the U.S.A. and U.K. may be approaching a peak. As it is a common ingredient of the diet, and one that may be consumed in comparatively large quantities, it is necessary to learn whether the consumption of sucrose has an effect on lipid metabolism different from that of other common dietary carbohydrates such as starch. The easiest method of attempting to study such effects is to take serial samples of fasting blood from subjects consuming diets containing various carbohydrates in isocaloric amounts. Estimations of serum lipid concentrations made on fasting blood are not, under these circumstances, influenced by the dietary intake of fat. When interpreting the results of such studies and applying them to lipid metabolism, it must be borne in mind that the sample of blood taken represents the balance between input and output at the time the sample was taken. Where lipid or other levels are raised, there is no means of assessing whether the raised level is due to an increase in input or decrease in output, or both. The effects of sucrose on blood lipid concentrations can be described, but in themselves are of limited academic or clinicial value; interpretation of the data so obtained must be made with reservations.

SOME EFFECTS OF DIETARY SUCROSE ON SERUM CHOLESTEROL LEVEL

Important dietary factors which influence the level of serum cholesterol are the amount of cholesterol in the diet[6] and the type of fat in the diet[7] and it is, therefore, difficult to assess the part played by dietary sucrose in the level of this lipid fraction. In rats, higher cholesterol levels were found when sucrose replaced starch in a diet containing cholesterol[8] and similar results were reported in rabbits.[9] From these and other studies it seems that in animals the effect of the type of carbohydrate only becomes apparent when cholesterol is present in the diet.

The effects of sucrose on the serum cholesterol level in man are not clear cut for reasons mentioned earlier. There are several reports which state that sucrose has no effect on the cholesterol level in man while others state that sucrose is associated with a slight rise in cholesterol concentration, when compared with other carbohydrates. Grande,[10] reviewed this topic and considered that '. . . the effect of dietary carbohydrate on serum cholesterol in normal man is very limited...'.

SOME EFFECTS OF DIETARY SUCROSE ON SERUM PHOSPHOLIPID LEVELS

As the serum phospholipid level does not seem to be implicated in the aetiology of any major clinical disorder, the study of this lipid fraction has not been so extensive as that of cholesterol or of triglycerides. It was reported that acute feeding of sucrose to rats resulted in a fall in the concentration of serum phospholipid.[11] In baboons, test meals of glucose were followed by a drop in the serum phospholipid level whereas sucrose meals were followed by a sharp rise in the phospholipid level.[12] A slight drop in the serum phospholipid level was found in patients with atherosclerosis after 10 days on a high sucrose diet.[13] Thus it seems that dietary sucrose has no consistent effect on serum phospholipid levels, though it may be of some importance that sucrose and other dietary carbohydrates have an effect on various phospholipid fractions. This has not been investigated.

SOME EFFECTS OF DIETARY SUCROSE ON THE SERUM TRIGLYCERIDE LEVELS

The question that has to be answered is whether sucrose is more or less effective than other dietary carbohydrates in raising the concentration of endogenous triglycerides in the fasting serum in those persons who have a carbohydrate-induced or Type IV hypertriglyceridemia.[14] Before presenting the evidence it must be pointed out that the concentration of triglycerides in the serum refers to the fasting state. It does not seem as if carbohydrates have any very striking effect on serum triglyceride levels (except a slight reduction, possibly due to insulin release) within a few hours of ingestion. It is not until 10–12 h after ingestion of carbohydrate that any marked changes in the fasting serum may be seen.

Experimental work on the effects of dietary sucrose on blood lipids is not extensive. The animal which has been largely used is the rat, but before embarking on dietary carbohydrate studies, careful consideration should be given to the fact that the lipid response to carbohydrate may be quite different depending on the strain of rat used.[15] Also in the rat, the absorption of fructose is quite different from man, in that after the intra-cellular hydrolysis of sucrose in the presence of sucrase to yield glucose and fructose, the fructose liberated diffuses back into the intestinal lumen and is absorbed lower down the gut.[16] A report on dogs in which dietary starch and sucrose were compared showed that in euthyroid animals there was no difference in the serum triglyceride response but in animals after thyroidectomy sucrose raised the triglyceride level.[17]

The baboon (Papio) has been used to study the effect of dietary sucrose on serum triglycerides. Results show that when male animals are given a liquid diet containing 75% (by weight) sucrose, with virtually no fat other than the fat soluble vitamins, the concentration of serum triglyceride rises during the first three weeks and then gradually falls so that by 17 weeks the levels are only slightly above the control value.[18] When a test meal of ^{14}C-U-sucrose is given after 13 weeks on the diet, the resulting specific activity in the serum glycerides is greater than that found after a comparable test meal given before the animals were put on the sucrose-rich diet. The triglyceride levels in the fasting serum of female baboons showed no response to the high sucrose diet, and further experiments revealed that natural oestrogens have the ability to prevent the rise in triglycerides associated with the high sucrose diet in the male baboons and that testosterone had no influence on this parameter in the female baboons.[19] Similar differences in metabolism of fructose and glucose have been found after intra-venous infusions of these monosaccharides[20] thus making unlikely such possibilities as rate of absorption and gut flora as complete explanations for the different lipid response to various dietary carbohydrates.

TRIGLYCERIDE LEVELS IN MAN

There have been several reports during the past few years of the effect that sucrose has on serum lipid level, especially triglyceride, in man, and these effects have been compared with those seen after giving comparable amounts of other dietary carbohydrates, notably starch. There are numerous papers which report that sucrose is associated with an increase in the fasting serum triglyceride concentration compared with other carbohydrates examined.[21–26]

There are also brief accounts which purport to show no difference between sucrose and other dietary carbohydrates on the fasting serum triglyceride level.[27, 28] In the majority of the experiments in man, the carbohydrate in the experimental diet has been given in an amount higher than that consumed in a normal diet, the reason being to discover whether such a stimulus provokes the metabolism in any way. The experimental diets are also relatively short term so that it may be that some adaptive changes have not been completed and consequently their effects not fully realised before terminating the experiment.

If sucrose can raise the level of triglycerides in the fasting serum to a greater extent than other dietary carbohydrates, then an obvious hypothesis to account for this finding is that it is the fructose moiety of sucrose which is responsible for the lipogenic effect of sucrose. It is not feasible to take fructose as the sole source of dietary carbohydrate as it produces an osmotic diarrhoea, but it is feasible to give approximately 200 g/day to an adult with no evidence of intestinal hurry. When this is done in man an increase in the fasting serum triglyceride concentration results, whereas glucose given in similar quantities has no such effect.[29]

Further evidence supporting the view that it is the fructose moiety of sucrose that is in some measure responsible for the lipid responses peculiar to sucrose, is found in the greater incorporation of ^{14}C fructose into serum triglyceride than is found with ^{14}C glucose, both in experimental animals and man.[12, 30]

If atherosclerosis and its complications are related in some way to lipid metabolism then the relatively low incidence of myocardial infarction in pre-menopausal women may indicate that the carbohydrate-lipid interrelationships in this group may be different from those in men. Experiments show that, whereas diets containing fructose raise the fasting serum triglyceride level in men, the reverse effect seems to occur in your women.[21, 29] Post-menopausal women respond to fructose-containing diets in a similar manner to men.

MECHANISM OF ACTION OF SUCROSE

If it is the fructose moiety that is responsible for the lipid response to sucrose being different from that of other dietary carbohydrates, then several hypotheses exist to account for the different metabolic handling of fructose. One of these possibilities is that fewer fructose molecules than glucose molecules may pass down the metabolic pathway to pyruvate and relatively more go to α-glycerol phosphate and thus the increase in triglyceride is due, not to increased formation of fatty acid, but to increased formation of the glycerol part of the triglyceride. If this were so, then an immediate precursor of α-glycerol phosphate should increase the triglycerides in men but not in young women. This has been found to be so, whether the glycerol is given as a single large dose or is given over a period of several weeks while on an otherwise normal diet.[32]

Further evidence to support the hypothesis that more ingested fructose than glucose goes into the glycerol part of the triglyceride molecule is seen in the incorporation of ^{14}C into the two parts of the triglyceride molecule after giving ^{14}C fructose and ^{14}C glucose by mouth. When this experiment is carried out in rats a greater proportion of the specific activity of triglycerides is found in the glycerol fraction of liver and serum triglycerides after fructose than after glucose.[33] In patients who have had myocardial infarcts the same pattern is seen.[34] It has been reported that the extent to which glucose forms fatty acids is dependent on the level of insulin[35, 36] and as fructose does not give rise to the same level of insulin secretion as glucose, this could be the explanation for this difference.

The effect of dietary sucrose on serum triglyceride levels is modified by other features of the diet. For example, if the lipid in the diet containing sucrose is cream, the fasting level of triglyceride increases, but if the cream is replaced by sunflower seed oil (high in polyunsaturated fatty acids) there is no change in the level of triglyceride.[37] Also the pattern of increase of fasting serum triglyceride level seems to depend on whether the sucrose is eaten in infrequent large quantities (gorging) or in frequent small quantities (nibbling).[38] Perhaps the type of protein in the sucrose-containing diet may influence the response to sucrose. These other factors which modify the effect of sucrose on blood lipids are very important, especially in terms of disease in man, because man does not eat sucrose as the sole carbohydrate, nor is the diet usually fat free, nor is the type of protein or meal-eating pattern constant.

Sucrose is able to influence fasting serum lipids, mainly triglycerides, when it is fed in large amounts, but whether under normal dietary circumstances this occurs or, if it does, whether it is important, is not yet known.

REFERENCES

1. AHRENS, E. H., HIRSCH, S., OETTLE, K., FARQUHAR, J. W. and STEIN, Y., *Trans. Ass. Am. Physns,* **74,** 134 (1961)
2. ALBRINK, M. J. and MAN, E. B., *Archs intern. Med.,* **103** (1959)
3. COHEN, A. M., BAVLY, S. and POZANSKI, R., *Lancet,* **2,** 1399 (1961)
4. TOGEL, O., BREZINA, E. and DURIG, A., *Biochem. Z.,* **1,** 296–345 (1913)
5. HIGGINS, H. L., *Am. J. Physiol.,* **41,** 258–265 (1961)
6. CONNOR, W. E., STONE, D. B. and HODGES, R. E., *J. clin. Invest.,* **43,** (1964)
7. BRONTE-STEWART, B., ANTONIS, A., EALES, L. and BROCK, J. F., *Lancet,* **2,** 521 (1956)
8. PORTMAN, O. W., LAWRY, E. Y. and BRUNO, D., *Proc. Soc. exp. Biol. Med.,* **91,** 321 (1956)
9. POLLOCK, O. J., *J. Am. Geriat. Soc.,* **9,** 349 (1961)
10. GRANDE, F., *Am. J. clin. Nutr.,* **20,** 176–184 (1967)
11. BRAGDON, J. H., HAVEL, R. J. and GORDON, R. S., *Am. J. Physiol.,* **189,** 63 (1957)
12. MACDONALD, I. and ROBERTS, J. B., *Metabolism,* **16,** 572 (1967)
13. PLESHKOV, A. M., *Terap. Arkh.,* **35,** 66 (1963)
14. LEVY, R. I. and FREDRICKSON, D. S., *Am. J. Cardiol.,* **22,** 576–583 (1968)
15. DURRAND, A. M. A., FISHER, M. and ADAMS, M., *Archs Path.,* **85,** 318–324 (1968)
16. DAHLQVIST, A. and THOMSON, J., *J. Physiol., Lond.,* **167,** 193–209 (1963)
17. SCHULTZ, A. L. and GRANDE, F., *J. Nutr.,* **94,** 71–73 (1968)
18. COLTART, T. M., *Ph.D. thesis,* University of London (1967)
19. COLTART, T. M. and MACDONALD, I., *Br. J. Nutr.* (In press, 1971)
20. JOURDAN, M. H., *Proc. Nutr. Soc.,* **28,** 63A (1969)
21. ANTAR, M. A. and OHLSON, M. A., *J. Nutr.,* **86,** 329 (1965)
22. GRANDE, F., ANDERSON, J. T. and KEYS, A., *J. Nutr.,* **86,** 313 (1965)
23. KAUFMANN, N. A., POZNANSKI, R., BLONDHEIM, S. H. and STEIN, Y., *Am. J. clin. Nutr.,* **18,** 261 (1966)
24. KUO, P. T. and BASSETT, D. R., *Am. intern. Med.,* **62,** 1199 (1965)

25. MACDONALD, I. and BRAITHWAITE, D. M., *Clin. Sci.*, **27,** 23 (1964)
26. McGANDY, R. B., HEGSTED, D. M., MYERS, M. L. and STARE, F. J., *Am. J. clin. Nutr.,* **18,** 237 (1966)
27. ANTONIS, A., ILES, C. and PILKINGTON, T. R. E., *Proc. Nutr. Soc.,* **27,** 2A (1968)
28. LEES, R. S., *Clin. Res.,* **12,** 549 (1965)
29. MACDONALD, I., *Am. J. clin. Nutr.,* **18,** 369 (1966)
30. MACDONALD, I., *Am. J. clin. Nutr.,* **21,** 1366 (1968)
31. BEVERIDGE, J. M. R., JAGANNATHAN, S. N. and CONNELL, W. F., *Can J. Biochem. Physiol.,* **42,** 999 (1964)
32. MACDONALD, I., *Br. J. Nutr.,* **24,** 537 (1970)
33. MARUHAMA, Y. and MACDONALD, I., *In preparation* (1970)
34. MARUHAMA, Y., *Metabolism* (in press, 1971)
35. CAHILL, G. F., LEBOEUF, B. and RENOLD, A. E., *J. biol. Chem.,* **234,** 2540 (1959)
36. VAUGHAN, M., *J. biol. Chem.,* **236,** 2196 (1961)
37. MACDONALD, I., *Am. J. clin. Nutr.,* **20,** 345 (1967)
38. MACDONALD, I., COLES, B. L., BRICE, J. and JOURDAN, M. H., *Br. J. Nutr.,* **24,** 413 (1970)

V

SUCROSE AND HUMAN DISEASE

Chapter Eighteen

ACUTE CLINICAL SYNDROMES RELATED TO SUCROSE INGESTION IN MAN

VALERIE BURKE
Royal Australasian College of Physicians Overseas Fellowship,
St. Bartholomew's Hospital, London

In man, symptoms follow within a few hours of sucrose ingestion in a group of uncommon disorders, most of which are caused by genetically determined or acquired enzyme defects. Sucrose itself may be responsible, but symptoms also result from disturbed metabolism of glucose or fructose formed by hydrolysis of sucrose *(Table 1).*

Table 1. SYNDROMES IN WHICH SYMPTOMS FOLLOW ACUTELY AFTER SUCROSE INGESTION

Symptoms caused by	*Disorder*
Sucrose	Sucrase-isomaltase deficiency Secondary disaccharidase deficiency
Glucose	Glucose-galactose malabsorption Secondary monosaccharide malabsorption
Fructose	Secondary monosaccharide malabsorption Fructosaemia Familial fructose and galactose intolerance

At the beginning of this century, ingestion of sugars was recognised as a cause of diarrhoea in some infants, and sugar free diets became popular in their management.[1–3] Howland, in 1921, postulated that an enzyme defect was responsible for diarrhoea caused by dietary sugars, but it was not until 1963 that intestinal disaccharidase deficiency was demonstrated in patients whose symptoms were related to sucrose ingestion.[5] The basic defect has not yet been characterised in the disorders of monosaccharide absorption although, in glucose-galactose malabsorption, the mechanism for transport of these sugars is affected. Deficiency of hepatic fructose-1-phosphate aldolase has been recognised since 1957 as the underlying abnormality in fructosaemia but the defect in familial fructose and galactose intolerance is unknown.

In modern Western communities, where carbohydrate constitutes 50–70% of the usual adult diet, the daily intake of sucrose averages about 80 g[6] (about 1·5 g/kg body weight). Young babies fed the common milk formulae sweetened with sucrose take approximately 15 g of sucrose daily (3 g/kg).

ABSORPTION OF SUCROSE

Normally there is negligible absorption of sucrose as the intact disaccharide, but if the intestinal mucosa is damaged[7] or sucrose concentration is sufficient,[8] small amounts of sucrose may be absorbed. Sucrose is not metabolised[9] and is excreted in the urine.

For normal absorption, sucrose must first be hydrolysed to its component monosaccharides, glucose and fructose, by the disaccharidase, sucrase, situated in the brush border of the small intestine.[10] The other disaccharidases of clinical importance are also found in the brush border and include the β-galactosidase lactase and the α-galactosidases sucrase, isomaltase and maltase. These are normally present in the approximate ratio of 1:2:2:5.[11] The number of disaccharidases isolated depends on the method of separation used[12,13] and, as yet, none of these proteins has been characterised.

Glucose released at the brush border by hydrolysis of sucrose is absorbed by means of a so far unidentified carrier mechanism shared by galactose.[14] Absorption of fructose is proportional to its concentration[15] but proceeds more rapidly than predicted from a process of simple diffusion. Significant conversion of fructose to glucose within the mucosa has been shown *in vitro*,[16] but is not an important pathway *in vivo*[15] and does not account for the rate of fructose absorption. Moreover, in glucose-galactose malabsorption, a rare disorder in which the carrier mechanism is impaired, fructose is absorbed more rapidly than glucose which is of similar molecular size. These observations suggest that fructose absorption is facilitated.

DISORDERS OF ABSORPTION

In all types of sugar intolerance caused by disorders of absorption, diarrhoea is the main symptom. The unabsorbed sugar, whether a disaccharide or monosaccharide, interferes with absorption of water from the small bowel because of its osmotic activity. Acid metabolites formed in the colon by bacterial fermentation contribute to diarrhoea by further increasing osmolarity and reducing pH, so impairing colonic absorption of water.[17] Watery stools are typical of disorders of sugar absorption.

PRIMARY AND SECONDARY SUCRASE DEFICIENCY

Sucrase deficiency may be an inherited disorder, always associated with isomaltase deficiency, or may be secondary to diseases such as gastro-enteritis which affect disaccharidase activity by damaging intestinal mucosa. Normally there is an enormous reserve of disaccharidase activity, more than adequate to hydrolyse the amount of disaccharide presented and although disaccharidase deficiency may be demonstrated in a sample of mucosa obtained by small bowel biopsy, symptoms do not always follow ingestion of the disaccharide in this secondary type of disaccharidase deficiency. In primary disaccharidase deficiency the enzyme defect is specific but in the secondary type all disaccharidases are usually affected to a varying degree and there may be evidence of other disturbances of absorption such as steatorrhoea. As the mucosa heals and normal villous architecture reappears, absorption returns to normal.

SUCRASE-ISOMALTASE DEFICIENCY

Diarrhoea following sucrose ingestion, presumably caused by sucrase deficiency, was first described in 1961.[18] Co-existing intolerance to isomaltose, a sugar made up of 2 glucose units joined by a 1-6 linkage, was later demonstrated.[19] Deficiency of sucrase and isomaltase in the intestinal mucosa of these patients was first shown by paper chromatography when their mucosal homogenates failed to hydrolyse sucrose and isomaltose.[5] Further reports confirmed these findings with Dahlqvist's quantitative assay; there was also a reduction in maltase activity as Dahlqvist[12] had predicted, since both enzymes also hydrolyse maltose. The consistent finding of both sucrase and isomaltase deficiency is explained by a defect in a structural gene affecting either a protein with more than one active site[20] or a polypeptide unit common to these enzymes. Recently, Eggermont[21] studying sedimentation coefficients of the α-glucosidases has supported this explanation by finding an abnormal sedimentation pattern in homozygotes with sucrase-isomaltase deficiency and, in one heterozygote, a combination of normal and abnormal patterns.

GENETICS

Prader and Auricchio[22] first suggested autosomal dominant inheritance, but subsequent studies[23] support an autosomal recessive trait by separating three groups among families with sucrase-isomaltase deficiency, expressing enzyme activity as the ratio of sucrase to lactase. Some of the siblings had normal ratios, others, like the probands, had very low values while all the parents and remaining sibs had activity intermediate between these two groups. These correspond to homozygous normals, patients homozygous for sucrase-isomaltase deficiency, and heterozygotes.

SYMPTOMS OF SUCRASE-ISOMALTASE DEFICIENCY

Patients have no symptoms until sucrose is introduced into the diet. Although isomaltase activity as well as sucrase activity is low, normal diets contain little isomaltose and very little is released by hydrolysis of starch so this is not clinically important.

Severity of symptoms depends on the patient's age, the amount of sucrose ingested and whether the patient is homozygous or heterozygous.

Diarrhoea is most dramatic in young homozygotes, particularly under the age of six months, fed with a sucrose-sweetened formula. They may lose as much as one litre of fluid daily in profuse diarrhoea which follows sucrose ingestion. Stools are typically frothy and watery; excoriation of the buttocks soon occurs but dehydration is the main clinical problem in these babies. If dietary sucrose is not recognised as the cause of diarrhoea, prolonged intravenous therapy becomes necessary and inadequate calorie intake leads to malnutrition and susceptibility to infection. Sucrase deficiency can be life-threatening in young babies.

Older children and adults rarely present with severe diarrhoea and dehydration is unusual. Bloating and abdominal cramps may follow sucrose ingestion and impaired growth and abdominal distension are usual. Such patients have usually first encountered sucrose after weaning, so avoiding the more dramatic effects of sucrose ingestion.

Although the difference in symptoms between babies and adults is partly explained by the lower dose of sucrose usually ingested, changes in bowel habit with colonic adaptation is probably also important.

At any age, the severity of diarrhoea is proportional to the dose of sucrose given. Heterozygotes may present with diarrhoea as neonates, but an amount of sucrose greater than that producing symptoms in homozygotes is necessary. Diarrhoea rarely occurs in heterozygotes beyond infancy unless very large amounts of sucrose are ingested.

Kerry and Townley[23] report one patient who is an exception to this pattern of symptoms. Discovered during family studies, this girl who was homozygous for sucrase-isomaltase deficiency had no symptoms even after a large dose of sucrose. They suggest that gastric acid may have hydrolysed sucrose sufficiently to prevent diarrhoea, but this intriguing observation is unexplained.

DIAGNOSIS

A detailed history relating the type of dietary sugar to the onset of diarrhoea is essential.

Tests for any type of sugar intolerance caused by impaired absorption of disaccharides or monosaccharides depend on demonstrating abnormal amounts of sugar and its breakdown products in the stools, on the effect of unabsorbed sugar on bowel contents, and on the insufficient rise in blood glucose after an oral load of sugar. Small bowel biopsy is necessary as the final step in diagnosis of some types of sugar intolerance.

EXAMINATION OF THE STOOLS

In children, the amount of sugar in the stool is readily estimated by a modification of the Clinitest method[24] provided that the specimen is collected on non-absorbent material so that the fluid part of the stool is tested. This method detects lactose, glucose, fructose and galactose but Clinitest will not react with sucrose which is not a reducing substance. Preliminary hydrolysis of sucrose by boiling with hydrochloric acid allows application of the Clinitest to the diagnosis of sucrose intolerance.[23] We have found the Clinitest to be reliable as a test for sugar intolerance in babies and children and useful in management as the stools become negative to Clinitest when dietary restriction of sugar is adequate. Other workers prefer to determine stool sugars chemically and identify them by chromatography.[25] Detection of stool sugars is a less reliable test for sugar intolerance in adults[26] possibly because of the smaller amount of sugar, in relation to body weight, ingested by adults.

Determination of lactic acid in the stools has also been used in diagnosis, a value of more than 50 mg per 100 g of stool being considered abnormal.[11] This test is not always reliable. In children whose intake of sugar has been restricted, there may be an interval of more than 24 hours after the test dose before fermentation products appear in the stools. Lactic acid values are even more variable in adults and no difference could be found between the lactic acid content of random stools from lactase deficient adults and controls.[26] Again the amount of dietary sugar for a normal adult probably affects this test but fermentation products may be absorbed in the colon. Stool lactic acid is no longer popular for diagnosis.

Changes in stool pH also depend on the presence of fermentation products and are even less reliable. Stool pH may fall below 5·5 in patients with uncomplicated gastro-enteritis; moreover, the stools may not be acid in patients with sugar intolerance.[11, 26]

OSMOTIC EFFECTS OF UNABSORBED SUGAR

Watery stools containing sugar following a dose of the appropriate carbohydrate are the most obvious manifestation of its osmotic effects and usually occur in children but not necessarily in adults. Laws and Neale[27] have used X-ray appearances to demonstrate changes in intestinal contents caused by unabsorbed sugar. Radio-opaque material mixed with 25 g of sugar is fed and follow-through studies show dilution of the opaque medium and rapid intestinal transit in patients with sugar intolerance.

SUGAR TOLERANCE TESTS

Tests measuring the rise in blood glucose after an oral dose of sugar are widely used. In adults, the usual dose given in 10% solution is 50 g and in children 2 mg/kg or 50 g/m^2 depending on the preference of the investigator.[11] In sucrase deficiency, the blood sugar rises less than 20 mg% after a test dose of sucrose but more than 30 mg% if the same amount of sugar is given half as glucose and half as fructose. The multiple factors influencing such oral tolerance tests are well summarised by Holdsworth[28] and explain the variable results sometimes found. However, such tests are useful in diagnosis and 80% accuracy has been claimed,[29] while Welsh[30] reports greater reliability using capillary blood. In many centres these tests are used for investigating children but, in our experience, the detection of abnormal amounts of sugar in the stools after a test dose of sugar is valuable and avoids the need for multiple blood samples.

Some workers prefer to combine these tests, collecting blood samples, stools and urine after a test dose of sugar. Disacchariduria is not diagnostic of disaccharide intolerance.

DISACCHARIDASE ASSAY IN SUCRASE-ISOMALTASE DEFICIENCY

Diagnosis finally depends on demonstrating reduced sucrase and isomaltase activity in histologically normal small-bowel mucosa. Disaccharidase activity is otherwise normal although maltase levels are slightly depressed. These findings distinguish sucrase-isomaltase deficiency from secondary disaccharidase deficiency, an important point in management since lifelong dietary modification is usually necessary in sucrase-isomaltase deficiency. Heterozygotes can be distinguished by disaccharidase assay and this is also of practical importance since their sucrose intake need not be so limited.

Although we recommend confirmation of the diagnosis by disaccharidase assay, small bowel biopsy has significant risks in young debilitated infants[31] and the methods previously described are useful in deciding rational dietary management while waiting for their general condition to improve.

MANAGEMENT

Replacement therapy with sucrase and isomaltase has been attempted[32] but sucrose restriction is simple and less expensive. Sucrose must usually be excluded from the diet for the first year of life, but, afterwards, some sucrose can be tolerated without producing symptoms. Restriction of starch is unnecessary since the small quantities of isomaltose presented in this way are insufficient to cause symptoms.

Patients' siblings should also be investigated and biopsy performed in all showing an abnormal response to oral sucrose. Older children with less florid symptoms will be detected and growth and stool pattern improved by sucrose restriction. Nursery staff should be warned to withhold sucrose-containing feeds from newborns in these families. The response to oral sucrose is assessed at about a month of age and, if sucrose intolerance seems likely, biopsy is deferred until the baby is at least six months older. Meanwhile sucrose should be avoided. Many of these newborns who develop diarrhoea after a test dose of sucrose will prove to be heterozygotes.

SECONDARY DISACCHARIDASE DEFICIENCY

Rarely, in infancy, sucrose intolerance follows gastroenteritis[33] or gastro-intestinal surgery in the newborn[34] but is much less common than lactose intolerance. Although disaccharidase deficiency occurs in other disorders such as coeliac disease,[35] in our experience ingestion of sucrose does not cause symptoms in these patients although rarely lactose may do so. We do not recommend routine exclusion of lactose and sucrose, as well as gluten, in the early management of coeliac disease which Arthur *et al.*[36] have suggested. In secondary disaccharidase deficiency, lactase, the most labile of these enzymes, is the first to reach a level at which symptoms follow lactose ingestion and, with continuing damage, there is sufficient reduction in sucrase activity to cause diarrhoea after feeding sucrose. Maltase deficiency is not clinically important. In these patients sucrase deficiency is always associated with lactase deficiency and the mucosa is usually histologically abnormal, in contrast to the primary group.

CLINICAL FEATURES

Secondary disaccharidase deficiency is likely when a young baby, recovering from surgery or gastro-enteritis, develops watery diarrhoea as he is graded from dilute milk feeds used during the acute illness to milk containing a higher concentration of sugar. Lactose intolerance is the usual cause of these symptoms and a lactose-free milk is substituted. Such formulae often contain sucrose, and diarrhoea continues in sucrose-intolerant infants. Diarrhoea is controlled by a feed containing no lactose or sucrose.

The tests previously described apply in the diagnosis of secondary disaccharidase deficiency and the stools contain abnormal amounts of sugar while lactose or sucrose are fed. Repeated small-bowel biopsy in such infants has shown return to normal enzyme activity and histology during a few weeks or months[33] and we do not recommend biopsy routinely in these infants. Tests of tolerance to lactose and sucrose should be repeated every one or two weeks until the response is normal. An unrestricted diet can then be recommended. Biopsy is indicated only when sugar intolerance continues for longer than 12 months and a persisting underlying disorder is suspected.

SECONDARY MONOSACCHARIDE MALABSORPTION

This poorly understood condition occurs only in young babies who are unable to tolerate the monosaccharides or more complex sugars contained in usual infant foods, including glucose, galactose and fructose. Disaccharidase activity and small bowel histology are normal. Monosaccharide malabsorption follows infections such as gastro-enteritis[37] and occurs after surgical relief of small bowel obstruction[34] in the newborn. In the surgical group, after resection of the obstructed area, stasis of intestinal contents occurs in the dilated bowel proximal to the site of anastomosis and there is overgrowth of faecal flora not normally present in the small bowel. Investigation of the 'non-surgical' group has also shown overgrowth of faecal flora in the upper small bowel but without an obvious site of stasis. In the 'contaminated bowel' syndrome abnormalities of small bowel flora are associated with changes in protein metabolism,[38] and in fat and vitamin absorption,[39] and perhaps some bacteria can also interfere with sugar absorption. It seems unlikely that the presence of unabsorbed sugar in the small bowel lumen itself promotes growth of faecal organisms, as these changes do not occur in patients with disaccharidase deficiency who also have unabsorbed sugar in the bowel lumen.[40]

CLINICAL FEATURES

Initially the method of presentation resembles that seen in secondary disaccharidase deficiency. However, diarrhoea continues when monosaccharides replace dietary disaccharides and diarrhoea is controlled only by limiting the intake of monosaccharides. Dietary modification is usually necessary for a few weeks and the gradual return to normal sugar tolerance correlates with the disappearance of small bowel organisms.

Diagnosis depends on demonstrating an abnormal response to test doses of glucose, galactose and fructose. Biopsy does not help in diagnosis, but in this puzzling condition culture of the small bowel contents is of interest to see whether the association of abnormal flora and monosaccharide malabsorption is invariable.

MANAGEMENT

An artificial feed is used, based on about 3% protein and 3% fat with appropriate vitamin and mineral supplements and to this glucose is added in the highest concentration tolerated. Usually a sugar-free feed is required for the first week and, as tolerance improves, glucose concentration can be increased at about 1% weekly until tolerance is normal. The usual milk feedings can then be reintroduced. Harries and Francis[41] have detailed dietary management with products obtainable in Great Britain. If abnormal small bowel flora are found in other cases, treatment with antibiotics, as used in the blind loop syndrome, may shorten the course of this disorder.

GLUCOSE-GALACTOSE MALABSORPTION

In 1962, the first reports of a congenital disorder of monosaccharide absorption appeared.[42] These rare patients are unable to absorb glucose, galactose or sugars containing them, while fructose is normally absorbed. Inheritance is autosomal recessive.[43]

Small-bowel biopsy shows normal histology and disaccharidase activity but, on incubation with C^{14} labelled glucose or galactose, there is defective uptake of these sugars.[44] Uptake of 3-ortho-methyl glucose, which shares the configuration necessary for the carrier mechanism, is also impaired and the defect is probably in this carrier. Glycosuria occurs in some of these patients suggesting that transport of glucose in the renal tubule is also impaired. Galactose does not share this mechanism which probably differs from the intestinal transport process.[21]

CLINICAL FEATURES

Diarrhoea begins soon after the newborn has the first sugar-containing feed. Sucrose causes less severe diarrhoea than other disaccharides because fructose is absorbed.

Diagnosis depends on sugar tolerance tests as previously discussed. Biopsy is not helpful in diagnosis and the autoradiographic methods showing impaired glucose uptake are really research tools.

MANAGEMENT

All sugars except fructose must be at first excluded and, in later life, considerably restricted. An artificial milk containing fructose as the only sugar is available as Galactomin, Trufood; otherwise the feeding is similar to that used for secondary monosaccharide malabsorption with the addition of fructose.

Newborn babies in families with glucose-galactose malabsorption must be carefully observed and a formula free of glucose and galactose should be used as soon as loose stools appear. Mortality is high if the condition is unrecognised.

DISORDERS OF FRUCTOSE METABOLISM

METABOLISM OF FRUCTOSE *(Figure 1)*

Fructose is principally phosphorylated to fructose-1-phosphate by fructokinase in liver and intestinal mucosa. Some fructose-6-phosphate forms in the presence of hexokinase

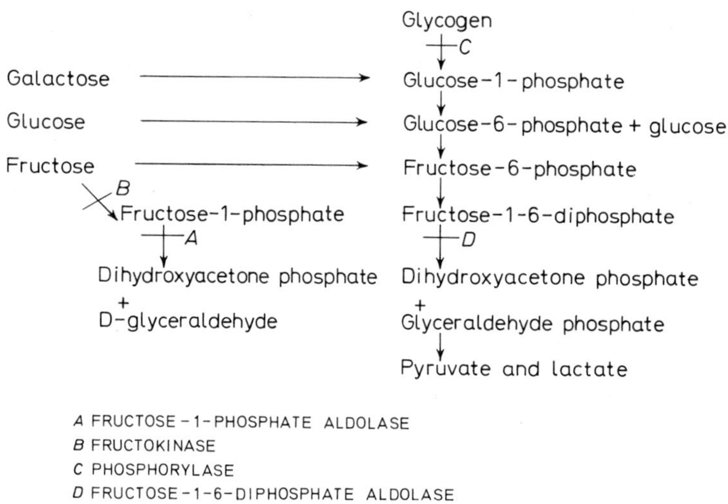

Figure 1. Summary of pathways for fructose metabolism showing steps affected in fructosaemia

which has greater affinity for glucose and this step is usually unimportant except in adipose tissue where glucose does not inhibit the reaction.[45] In liver and intestine, fructose-1-phosphate aldolase splits fructose-1-phosphate to 2 trioses, D-glyceraldehyde and dihydroxyacetone phosphate, which may then enter the glycolytic pathway. Fructose-6-phosphate and fructose-1-6-diphosphate, formed in the presence of phospho-fructo-kinase and ATP, are both important intermediates in glycolysis. Fructose-1-6-diphosphate aldolase then splits the diphosphate to dihydroxyacetone phosphate and glyceraldehyde phosphate. Glucose, glycogen and galactose share this glycolytic pathway which is reversible for gluconeogenesis in the presence of ATP.

FRUCTOSAEMIA

Deficiency of hepatic fructose-1-phosphate aldolase is the cause of this disorder and is associated with a variable decrease in fructose-1-6-diphosphate aldolase activity.[46–49] There are a number of explanations for the apparent deficiency of two enzymes. Fructose-1-phosphate, which accumulates in fructosaemia, inhibits fructose-1-6-diphosphate aldolase[47] and fructose-1-phosphate aldolase has appreciable diphosphoaldolase activity.[50] However, Levin *et al.*[49] consider two different explanations. In some cases, where the ratio of diphosphate to phosphate aldolase approaches normal (1·5:1), a defect in a control gene causing decreased amounts of structurally normal enzymes explains the observations. When the ratio is high (>5:1) a mutation in a structural gene producing an abnormal polypeptide with reduced enzyme activity is more likely.

In the absence of fructose-1-phosphate aldolase (*A*), fructose-1-phosphate accumulates, inhibiting fructokinase (*B*), and blood fructose rises. The sequestration of phosphate in this complex leads to decreased inorganic phosphorus and ATP. Serum magnesium rises because magnesium is usually combined with ATP intracellularly and leaks out of the cell when ATP is deficient.[48] If exposure to fructose continues, hypomagnesaemia may result from renal losses of magnesium entering the serum in this way. Hypoglycaemia, which is the most usual feature in fructosaemia, is not completely explained, but probably results from both impaired glycogenolysis and gluconeogenesis. Insulin levels and peripheral glucose utlisation are normal.[51, 52] Defective glycogenolysis results from deficiency of inorganic phosphorus and AMP which converts phosphorylase (*C*) from the inactive '*a*' to the active '*b*' form. Decreased activity of fructose-1-6-diphosphate aldolase (*D*), together with depletion of ATP, leads to impaired gluconeogenesis. Pyruvate and lactate increase after fructose administration because of the defect in this pathway.

Hepatic and renal damage occur in young babies with fructosaemia. Deficiency of intracellular ATP and inorganic phosphorus interferes with normal cell metabolism. Elevation of serum transaminase, an index of hepatocellular damage, occurs as early as 1½ hours after fructose and quickly reverts to normal when dietary fructose is withdrawn.[48]

The renal abnormality is manifest by tubular dysfunction with amino-aciduria, glycosuria, hyperphosphaturia, hypokalaemia and metabolic acidosis (Fanconi syndrome). This is also reversible after fructose restriction.

If serum fructose is sufficiently elevated some fructosuria occurs, but this is inconstant.

Activity of both aldolases has been studied in specimens of small bowel mucosa. Froesch *et al.*[47] reported reduced fructose-1-phosphate aldolase and normal diphosphoaldolase activity in a single patient but Levin *et al.*[49] found, in their patient, no reduction of either enzyme in intestinal mucosa whereas both hepatic enzymes were deficient.

INHERITANCE

Most cases seem consistent with autosomal recessive inheritance, but three children each with one affected parent have been reported. Possibly these parents are heterozygotes but Cornblath[51] suggests dominant inheritance to explain the incidence in a family he describes. Heterozygotes cannot, as yet, be identified so the mode of inheritance remains unsettled.

SYMPTOMS

Severity of symptoms depends on the age at which fructose is first introduced. Young infants given a formula sweetened with sucrose may present with vomiting and poor weight gain. Jaundice, liver enlargement and, more rarely, ascites, occur and if fructose administration continues cirrhosis follows. Hypoglycaemia may be the presenting feature with convulsions, loss of consciousness or weakness and sweating. There may be a bleeding tendency, usually related to prothrombin deficiency, but one reported case had afibrinogenaemia. Renal tubular dysfunction leads to severe acidosis with biochemical abnormalities typical of Fanconi's syndrome which occurs in many other disorders, and fructosaemia may escape diagnosis while other possibilities are investigated.

Young infants may succumb to the disease if the relationship between symptoms and fructose ingestion is not recognised.[49]

Older children and adults with fructosaemia have usually first encountered fructose at the time of weaning and symptoms are less dramatic, probably because they develop an aversion to sweet foods and so ingest very little fructose. Interestingly, most of these patients have been free of caries.[47] Many such cases have been recognised during family studies and there may have been no symptoms other than difficulty in weaning because of dislike of fructose-containing foods. However nausea and vomiting may follow fructose ingestion, or hypoglycaemia may be the presenting feature.

Symptoms are similar whether fructose is given intravenously or orally but nausea is less troublesome after an intravenous dose. This may be explained by deficiency of intestinal fructose-1-phosphate aldolase as found in Froesch's patient; Levin's patient with normal intestinal aldolase activity did not completely reject sweet foods in contrast to other patients.

DIAGNOSIS

Fructosaemia has many manifestations, each of which occurs in numerous other disorders. The relationship to fructose ingestion is easily overlooked, perhaps with fatal outcome. If a seriously ill young baby fed with a sucrose-sweetened formula presents such features, sucrose-free feedings must be substituted without waiting for laboratory confirmation of the diagnosis.

Diagnosis depends on the characteristic biochemical changes following administration of fructose. Froesch[46] recommends this should be given intravenously to avoid nausea and vomiting after an oral dose and uses 0·25 g/kg in adults or 3 g/m^2 in children. In small infants with liver damage this dose should be doubled or trebled before the diagnosis is discarded, as one such infant reported by Froesch failed to show the typical response to 3 g/m^2 of fructose. Levin[49] uses 1 g/kg for oral fructose tolerance tests in children.

After the administration of fructose, serum inorganic phosphorus falls. This provides the best guide to diagnosis and although blood glucose usually also falls, the change in phosphate levels may be the only sign. Blood fructose sometimes rises, but this response is too inconstant to use in diagnosis.

In patients showing the typical response to fructose, diagnosis should be confirmed by assaying hepatic fructose-1-phosphate aldolase activity. Results vary between 0 and 16% of normal; fructose-1-6-diphosphate aldolase is even more variable ranging from less than 1 to 87% of normal.[46] Intestinal mucosal enzymes are apparently not consistently decreased and although it would be interesting to document intestinal aldolase activity in more patients, this cannot be relied on for diagnosis.

MANAGEMENT

All the abnormalities are reversed by a fructose-free diet and mental development and growth are normal. Older patients will usually have devised their own fructose-free diet.

In a family known to have fructosaemia, newborns should be investigated for this disorder soon after birth. Meanwhile, milks sweetened with sucrose must be avoided.

FAMILIAL FRUCTOSE AND GALACTOSE INTOLERANCE

This is an extremely rare disorder, only one family with two affected sisters having been described.[53] Hypoglycaemia was the main feature in one patient, while her sister showed biochemical changes but no symptoms after administration of fructose or galactose. Both differed from patients with fructosaemia in showing no distaste for sweet foods.

Severe hypoglycaemia followed fructose administration and, unlike fructosaemia, there was a slow fall in blood glucose after galactose. Inorganic phosphate fell after fructose just as in fructosaemia. Erythrocyte galactose-1-phosphate uridyl transferase was normal, which excluded galactosaemia; liver enzymes were not investigated. Very high fasting levels of insulin were found in both patients,[52] but fell during hypoglycaemia. Hyperinsulinism also occurred after oral glucose but the blood sugar curves were normal.

There are obvious differences between this disorder and fructosaemia but the basic defect remains to be elucidated.

REFERENCES

1. FINKELSTEIN, H. and MEYER, L. F., *Münch. med. Wschr.*, **58**, 340 (1911)
2. GRULEE, C. G., *J.-Lancet*, **34**, 141 (1912)
3. SCHMIDT, A. and STRASBERGER, J., *Dt. Arch. klin. Med.*, **69**, 570 (1901)
4. HOWLAND, J., *Trans. Am. pediat. Soc.*, **33**, 11 (1921)
5. ANDERSON, C. M., MESSER, M., TOWNLEY, R. R. W. and FREEMAN, M., *Pediatrics*, **31**, 1003 (1963)
6. NEALE, G., *Hosp. Med.*, **2**, 1372 (1968)
7. GRYBOSKI, J. D., THAYER, W. R. JR., GABRIELSON, I. W. and SPIRO, H. M., *Gastroenterology*, **45**, 633 (1963)
8. FORDTRAN, J. S. and INGELFINGER, F. J., In: *Handbook of Physiology. Sect. 6. Alimentary Canal. Vol. 3. Intestinal absorption.* Ed. by C. F. Code, Am. Physiol. Soc., Washington D.C., 1457 (1968)
9. WESER, E. and SLEISENGER, M. H., *J. clin. Invest.*, **46**, 499 (1967)
10. JOHNSON, C. F., *Science, N.Y.*, **155**, 1670 (1967)
11. HOLZEL, A., *Archs Dis. Childh.*, **42**, 341 (1967)
12. DAHLQVIST, A., *J. clin. Invest.*, **41**, 463 (1962)
13. SEMENZA, G., AURICCHIO, S. and RUBINO, A., *Biochim. biophys. Acta*, **96**, 487 (1965).
14. CRANE, R. K., 'Absorption of sugars', In: *Handbook of Physiology, Sect. 6. Alimentary Canal, Vol. 3. Intestinal Absorption,* Ed. by C. F. Code. Am. Physiol. Soc., Washington D.C. 1323 (1968)
15. HOLDSWORTH, C. D. and DAWSON, A. M., *Clin. Sci.*, **27**, 371 (1964)
16. WHITE, L. W. and LANDAU, B. R., *J. clin. Invest.*, **44**, 1200 (1965)
17. CHRISTOPHER, N. L. and BAYLESS, J. M. *Abstr. Gastroenterology.* **56**, 1250 (1969)
18. WEIJERS, H. A., VAN DE KAMER, J. H., DICKE, W. K. and IJSSELING, J. *Acta paediat., Stockh.*, **50**, 55 (1961)
19. AURICCHIO, S., DAHLQVIST, A., MURSET, G. and PRADER, A., *Lancet*, **i**, 1303 (1962)
20. LAUNIALA, K., PERHEENTUPA, J., VISAKORPI, J. and HALLMAN, N. *Pediatrics*, **34**, 615 (1964)
21. EGGERMONT, E., *Thesis*. University Catholique de Louvain, Belgium (1968)
22. PRADER, A. and AURICCHIO, S., *A. Rev. Med.*, **16**, 345 (1965)
23. KERRY, K. R. and TOWNLEY, R. R. W., *Aust. paediat. J.*, **1**, 223 (1965)
24. KERRY, K. R. and ANDERSON, C. M., *Lancet*, **i**, 821 (1964)
25. ABRAHAM, J. M., LEVIN, B., OBERHOLZER, V. G. and RUSSELL, A., *Archs Dis. Childh.*, **42**, 592 (1967)
26. McMICHAEL, H. B., WEBB, J. and DAWSON, A. M., *Lancet*, **i**, 717 (1965)
27. LAWS, J. W. and NEALE, G., *Lancet*, **ii**, 139 (1966)
28. HOLDSWORTH, C. D., *Gut.* **10**, 422 (1969)
29. NEWCOMER, A. D. and McGILL, D. B., *Gastroenterology*, **53**, 371 (1967)
30. WELSH, J. D., *Gastroenterology*, **51**, 445 (1966)
31. SHWACHMAN, H., KHAW, K. T. and ANTONOWICZ, I., *Pediatrics*, **43**, 460 (1969)
32. WEIJERS, H. A., VAN DE KAMER, J. H., MOSSEL, D. A. A. and DICKE, W. K., In: *Disorders due to intestinal defective carbohydrate digestion and absorption,* Ed. by Durand, P., Pensiero Scientificó, Rome (1964)
33. BURKE, V., KERRY, K. R. and ANDERSON, C. M., *Aust. paediat. J.*, **1**, 147 (1965)
34. BURKE, V. and ANDERSON, C. M., *Aust. paediat. J.*, **2**, 219 (1966)
35. HAEMMERLI, V. P., KISTLER, H. J., AMMAMM, R., MARTHALER, T., SEMENZA, G., AURICCHIO, S. and PRADER, A., *Am. J. Med.*, **38**, 7 (1965)

36. ARTHUR, A. B., CLAYTON, B., COTTOM, D. G., SEAKINS, J. W. T. and PLATT, J. W., *Lancet,* **i,** 172 (1966)
37. BURKE, V. and DANKS, D. M., *Lancet,* **i,** 1177 (1966)
38. JONES, E. A., CRAIGIE, A., TAVILL, A. S., FRANGLEN, G. and ROSENOER, V. M., *Gut,* **9,** 466 (1968)
39. TABAQCHALI, S. and BOOTH, C. C., *Br. med. Bull.,* **23,** 285 (1967)
40. GRACEY, M., BURKE, V. and ANDERSON, C. M., *Lancet,* ii, 384 (1969)
41. HARRIES, J. T. and FRANCIS, D. E. M., *Acta paediat., Stockh.,* **57,** 505 (1968)
42. LINDQVIST, B. and MEEUWISSE, G. W., *Acta paediat., Stockh.,* **51,** 674 (1962)
43. MELIN, K. and MEEUWISSE, G. W., *Acta paediat., Stockh. (Suppl.),* **188,** 19 (1969)
44. SCHNEIDER, A. J., KINTER, W. B. and STIRLING, C. E., *New Engl. J. Med.,* **274,** 305 (1966)
45. SCHAPIRA, F., SCHAPIRA, G. and DREYFUS, J. C., *Enzymol. biol. Chim.,* **1,** 170 (1962)
46. FROESCH, E. R., 'Essential fructosuria and hereditary fructose intolerance'. In: *The Metabolic Basis of Inherited Disease,* 2nd edn, Ed. by Stanburg, J. B., Wyngaarden, J. B. and Fredricksen, D. S., McGraw Hill Book Co. Inc., New York, 133 (1966)
47. FROESCH, E. R., WOLF, H. P., BAITSCH, H., PRADER, A. and LABHART, A., *Am. J. Med.,* **34,** 151 (1963)
48. LEVIN, B., OBERHOLZER, V. G., SNODGRASS, G. J. A. I., STIMMLER, L. and WILMERS, M. J., *Archs Dis. Childh.,* **38,** 220 (1963)
49. LEVIN, B., SNODGRASS, G. J. A. I., OBERHOLZER, V. G., BURGESS, E. A. and DOBBS, R. H., *Am. J. Med.,* **45,** 826 (1968)
50. KALETTA-GINNENDER, V., WOLF, H. P. and LEUTHARDT, F., *Helv. Chim. Acta,* **40,** 1027 (1957)
51. CORNBLATH, M., ROSENTHAL, I. M., REISNER, S. H., WYBREGT, S. A. and CRANE, R. K., *New Engl. J. Med.,* **269,** 1271 (1963)
52. SAMOLS, E. and DORMANDY, T. L., *Lancet,* **i,** 478 (1963)
53. DORMANDY, T. L. and PORTER, R. J., *Lancet,* **i,** 1189 (1961)

Chapter Nineteen

SUCROSE AND DENTAL CARIES

R. L. HARTLES
Professor of Dental Science, School of Dental Surgery, University of Liverpool

Dental caries is a disease that seems to increase rather than decrease with improvements in the standard of living.

The initiation of the carious lesion is a surface phenomenon occurring at the interface between the enamel surface and the base of the dental plaque. The challenge to the integrity of the enamel surface is compounded of dietary and microbiological factors with considerable interaction between the two. Any explanation of the aetiology must take into account the fact that the attack on the tooth surface is not indiscriminate but occurs at certain localised points where there is an accumulation of dental plaque. The plaque consists of an adherent mass of micro-organisms, salivary degradation products, the products of microbial metabolism and small amounts of dietary debris. It occurs in greatest quantity in those sites on the tooth surface which are least self-cleansing —for example, in the pits and fissures of the occlusal surfaces and between the teeth. The organisms in the dental plaque can convert fermentable carbohydrates to organic acids, and it is these acids that by lowering the pH locally cause the initial breach in the highly mineralised enamel.

There are, therefore, three major factors required for the production of dental caries:

(1) A tooth surface
(2) A dental plaque
(3) Substrates for use by the plaque micro-organisms.

The factors controlling the onset of the disease are primarily of environmental origin, but the situation may be influenced by the physical and chemical nature of the tooth surface and the anatomical arrangement of the teeth in the jaw. Thus it is probable that a fluoridated tooth surface will resist a given cariogenic challenge to a greater degree than a similar non-fluoridated tooth surface. However, if the challenge provided by the environment is sufficiently great it is likely that any tooth surface will succumb to the disease.

It is doubtful whether the term caries 'immune' should be used either in relation to a tooth or a person, but it is necessary to keep an open mind on this matter at the present time. In the present state of knowledge it is wise to avoid the use of the term 'immunity' in relation to dental caries.

The use of the term 'caries-susceptible' is admissible in defined circumstances. Thus, other things being equal, a tooth surface with a low fluoride content may be described as more 'susceptible' to caries than one with a high fluoride content. On the other hand approximal enamel may become carious more readily than buccal or lingual enamel without being more 'susceptible', because the challenge is different. The potentially caries-active areas are those that most readily accumulate plaque, the caries-inactive areas are those on which plaque does not easily form.

It is the interrelationship between plaque, saliva and dietary carbohydrates which is so important in the initiation of the carious lesion.

The two major sources of carbohydrate in contemporary Western diets are starches of various kinds and sucrose. The starches of high molecular weight are not immediately available substrates for micro-organisms; it is necessary that they be hydrolysed to smaller units by amylase. This can be done but it takes time. The small sucrose molecule on the other hand provides an immediate source of substrate for many oral bacteria.

CLINICAL EVIDENCE CONCERNING SUCROSE AND DENTAL CARIES

During the last war and the immediate post-war years, there was, in several countries, a reduction in the availability of sugar and refined carbohydrate. In this country, the major changes in the dietary consumption of carbohydrates were qualitative rather than quantitative. Less sugar and highly refined flour were eaten, but the consumption of high-extraction flours, potatoes and vegetables was increased. Total carbohydrate intake was not significantly altered. The interesting fact emerged that in the United Kingdom,[1, 2, 3] Scandinavia[4] and Japan[5] there was a progressive decrease in the prevalence of caries in young children. This decrease was observed until 1947–48 when, with the gradual relaxation of restrictions and the increase in sugar consumption, the trend was reversed. It was suggested[6] that in this country, the improvement in dental health was related to an increased consumption of protective foods. However, a similar improvement was observed in Finland[7] where the general war-time standard of nutrition was poor. The most likely explanation for the improvement in dental health lay in the non-availability of sugar, especially for the manufacture of confectionery for between-meal eating.

There is other indirect evidence relating sugar consumption to dental caries, arising from observations of the impact of Western diets on less sophisticated peoples. Caries in the Eskimo increased when they adopted a 'civilised diet', the major change being an increase in sugar consumption.[8]

Direct evidence of an association between sugar consumption and caries was provided some years ago by a study on institutional patients at Vipeholm in Sweden.[9] The findings indicated that an increased consumption of sugar could cause an increase in caries, but the effect depended not only on the quantity of sugar consumed but on the manner of its administration. When the extra sugar was taken in solution it had little or no effect on subsequent caries. The greatest increase occurred in patients who consumed sugary confections between meals, and in this respect toffees and caramels were more cariogenic than chocolate. Thus the important finding was that there was a direct correlation between an increase in caries experience and the consumption of sugar between meals. A study was carried out in the United Kingdom, under the auspices of the Medical Research Council, whilst sugar rationing was still in force. Children in institutional homes were given an increased sugar ration corresponding to the pre-1939 consumption. The sugar was consumed only at meal times and no effect was observed, after two years, on the development of new caries.[10]

Thus from all the clinical studies there is enough evidence to establish a prima facie case that sucrose is potentially the dietary component most harmful to the teeth especially when consumed between meals.

These findings, however, do not permit any conclusions to be drawn concerning the specific cariogenicity of sucrose. They merely indicate that a simple sugar such as sucrose is cariogenic.

It is now necessary to ask the question, does sucrose assume importance in the promotion of dental caries merely because it happens to be the only common sugar consumed in large quantity, or has it specific properties which render it more harmful than other sugars? If the former is the case, reduction in sucrose consumption should reduce dental caries, but substitution of, say, glucose for sucrose would have no beneficial effect. If the latter is true replacement of sucrose by other sugars might be beneficial.

CARIES IN EXPERIMENTAL ANIMALS

It is not morally justifiable to seek further evidence concerning the relative cariogenicity of various carbohydrates by deliberately attempting to increase caries in young children by dietary means. The only way to investigate this problem in a positive fashion under controlled conditions is to use experimental animals maintained on carefully defined diets.

Certain strains of rat and hamster are susceptible to dental caries and the pathology of its development is similar to that in man, except that the time scale is much reduced.

In these animals it has been shown that diets containing 70% of uncooked starches derived from wheat, maize or potato are virtually non-cariogenic.[11, 12] Diets containing 70% sucrose are highly cariogenic. Heat-treated products such as roll-dried maize or wheat starch do cause some caries. Manufactured products such as biscuits also cause caries. Other sugars, e.g. glucose, fructose, maltose, galactose and lactose are all highly cariogenic.[13, 14] There is evidence, however, from several workers that sucrose is slightly but significantly more cariogenic than other sugars. It is only fair to say, however, that in some strains of rat this difference cannot be demonstrated.[15]

Thus it can be concluded that when diets contain 70% of one of the common sugars they are all highly cariogenic, but sucrose is in some conditions the most cariogenic.

Since caries begins beneath a dental plaque, it becomes of first importance to investigate the effect of different dietary carbohydrates on the microbial metabolism occurring within the plaque.

The ability of the plaque micro-organisms to convert many simple sugars with equal ease into acidic end-products is well established. 70% of the plaque micro-flora are streptococci of one sort or another. All of these bacteria can convert sugars into acids. It would seem probable therefore that ease of conversion to acids is not the only factor concerned in the cariogenicity of a carbohydrate. Nor is it likely that the ability to convert carbohydrate to acid is the only microbial activity associated with caries. Some oral streptococci have the ability to synthesise polysaccharide. Before discussing polysaccharide synthesis, it would be well to consider briefly the composition and formation of plaque.

The plaque consists of a matrix in which are distributed the oral flora and other cellular material. Two of the major components of the plaque matrix are protein and carbohydrate. The protein is almost certainly derived from the saliva, the carbohydrate is certainly not.

Saliva contains glycoproteins (mucins) and it was thought that these glycoproteins were precipitated on the tooth surface perhaps by a local change in pH. However, the characteristic carbohydrate components of salivary glycoprotein are sialic acid (N. acetylneuraminic acid) and fucose, neither of which can be identified in plaque. Leach[16] has suggested, therefore, that the origin of the plaque protein is from salivary glycoprotein that has lost its sialic acid by the action of the enzyme neuraminidase (sialase). The molecule will be maintained in its extended soluble form by the self-repelling effect of the negative charges. If these charges are removed the molecule will collapse, assume a globular structure and be precipitated. This can occur in two ways: either

(1) by a fall in pH which will suppress the ionisation of the sialic acid, or
(2) by enzymic removal of the sialic acid by neuraminidase.

Leach believes the second mechanism to be the operative one, since the plaque protein has a similar composition to the protein part of the salivary glycoprotein, minus the carbohydrate. We cannot, however, entirely exclude the possibility that the sialic acid and fucose could be removed enzymically after precipitation.

It is now widely agreed that the carbohydrate in the plaque is of microbial origin.

PLAQUE POLYSACCHARIDES

The polysaccharides in the plaque can be classified under two heads:

(*a*) intracellular polysaccharides
(*b*) extracellular polysaccharides

Both are produced enzymically from simple sugars by the plaque microbiota.

INTRACELLULAR POLYSACCHARIDES

Gibbons and Socransky[17] have studied the formation of intracellular iodophyllic polysaccharides of the amylopectin or glycogen type by oral micro-organisms and by the dental plaque. Various species cultivated from plaque—streptococci, fusiforms, diphtheroids—were found to produce large amounts of intracellular polysaccharide. This iodine-staining amylopectin could be demonstrated in organisms grown on glucose, maltose or sucrose. Samples of human plaque material was also shown to contain micro-organisms with intracellular iodine-staining polysaccharide.

Further, studies on caries plaques showed a greater proportion (60%) of cultivatable amylopectin formers than could be obtained from plaque from carious-free people (13%).

Studies on gnotobiotic animals[18, 19] showed that cariogenic strains of streptococci were strong producers of amylopectin; caries inactive strains were weak producers of amylopectin.

It has also been shown that the intracellular polysaccharides can be readily degraded to organic acids when exogenous sources of carbohydrates are not available for energy purposes. So intracellular polysaccharides could be formed by the plaque micro-organisms when exogenous carbohydrate is abundant and used up during periods of scarcity, thus prolonging the period when the plaque is at a low pH. In passing, it is of interest to note that amylopectin synthesis is inhibited by about 50% by 10 p.p.m. F at pH 7·2; at pH 5·5, 1–2 p.p.m. F causes a 30% inhibition.[20] So, in the intracellular production of polysaccharide, there is a facet of plaque metabolism of possible importance in the aetiology of dental caries.

EXTRACELLULAR POLYSACCHARIDE

Certain oral bacteria can produce a mucinous extracellular polysaccharide.[21, 22, 23, 24]

Studies have been carried out on strains of oral streptococci described as cariogenic, i.e. they produce caries in gnotobiotic animals. These organisms when grown in media containing sucrose produce large amounts of extracellular polysaccharides which are largely polymers of glucose, with a smaller amount of a fructose polymer.[25] In similar studies with non-cariogenic streptococci it was claimed that the proportion was reversed, i.e. there is more fructose polymer than glucose polymer. These polysaccharides are of the dextran type (glucose) or levan type (fructose).

In vitro 'cariogenic' streptococci produce more adherent polysaccharide than do 'non-cariogenic' streptococci when grown in the presence of sucrose.

Most of the known cariogenic streptococci produce 'dextran' from sucrose *in vitro*, but not in any great quantity from other sugars. This suggests that in these organisms the dextran-sucrose enzyme system is involved, which demands the presence of sucrose as a substrate for the formation of 1:6 linked dextrans.

Thus, *in vitro* it can be shown that certain oral streptococci produce 'dextran', others 'levan' (the former are associated with cariogenic, the latter with non-cariogenic organisms, but the division is not absolute).

However, studies on plaque show that the polysaccharide accumulating *in vivo* is predominantly of the dextran type.[26] Further it has been shown that levans are readily hydrolysed by the oral flora whereas dextrans are not.

Hence, any process leading to the synthesis of dextrans will help to build up the bulk of the plaque since they readily form adherent precipitates in the presence of saliva. Polysaccharide synthesis may well prove to be of great importance in the mechanism of plaque formation and accumulation. It is, however, not justifiable to claim that all plaque polysaccharide formation is dependent on dextran synthesis from sucrose; other mechanisms must operate.

POLYSACCHARIDE SYNTHESIS, PLAQUE AND CARIES IN ANIMALS

In the molar teeth of rodents, caries develops in two main sites, in the depths of the occlusal fissures and on the normally self-cleansing buccal and lingual surfaces. These latter smooth-surface lesions occur only when there is an accumulation of dental plaque. Occlusal caries is less dependent on plaque adherence since micro-organisms are readily impacted in the fissures. Thus factors disposing towards heavy plaque formation will be more likely to affect the development of smooth surface lesions than occlusal lesions.

It has been shown in animals whose indigenous oral flora has been suppressed, and whose mouths have been implanted with a cariogenic streptococcus, that heavy deposits of plaque containing polysaccharide are readily formed when the diet contains sucrose but not when it contains glucose, and significantly more caries is found when sucrose is present.[27, 28] This is an artificial situation because the organism concerned does not thrive on glucose, but it does show that a plaque containing much polysaccharide is highly cariogenic. However, it is not valid to imply that all caries and all plaque formation is dependent on polysaccharide synthesised from sucrose, since caries can be produced in fissures and in smooth surfaces in rats receiving diets containing sugars other than sucrose.

EFFECT OF CONCENTRATION OF SUCROSE ON LEVAN SYNTHESIS

Plaque can produce both dextran and levan from sucrose. Normally *in vivo* there is much less levan present because it is enzymically more labile than is dextran.

The effect of increasing concentration of sucrose shows that at low concentrations more dextran than levan is found and at 15% sucrose more levan than dextran.[29]

Similarly, rinsing the mouth with 10% sucrose for 2 and 10 minutes shows an increase in levan production which falls away with time.

So there are four factors of plaque metabolism of carbohydrate that may prove of importance in the causation of dental caries.

(1) Acid production—non-specific both in relation to organisms and to substrates.

(2) Intracellular synthesis of polysaccharide of the glycogen amylopectin type—non-specific in relation to substrate, partial specificity in relation to bacteria. Acts as food reserve for bacteria.

(3) Extracellular synthesis of dextran by dextran sucrase—specific for sucrose—selectively specific for bacteria. Dextran

is relatively stable and accumulates in plaque. Adds to physical bulk of plaque.

(4) Extracellular synthesis of levans by levan sucrase—specific for sucrose and some micro-organisms. Levan relatively labile—can act as food reserve.

EFFECT OF CONCENTRATION OF SUCROSE IN DIET ON DENTAL CARIES IN THE RAT

Until the percentage of sucrose in the diet is reduced below 25% there is no decrease in caries. The maximum levan production occurs *in vitro* at a concentration of 15% sucrose. It is not unreasonable to suggest that there may be a relationship between these two observations; namely that caries increases with increasing dietary sucrose content until the level is reached which ensures maximum levan production.

REFERENCES

1. MELLANBY, M. and MELLANBY, H., *Br. med. J.*, ii, 409 (1948)
2. MELLANBY, H. and MELLANBY, M., *Br. med. J.*, i, 1341 (1950)
3. MELLANBY, M. and MELLANBY, H., *Br. med. J.*, i, 51 (1951)
4. TOVERUD, G., *Proc. R. Soc. Med.*, **42**, 249 (1949)
5. TAKEUCHI, M., *Int. dent. J., Lond.*, **11**, 463 (1961)
6. MELLANBY, M. and COUMOULUS, H., *Br. med. J.*, i, 837 (1944)
7. WILSKA, A., *Odont. Tidskr.*, **54**, 296 (1946)
8. PEDERSON, P. O., *Dt. Zahn- Mund- u. Kieferheilk*, **4**, 728 (1939)
9. GUSTAFSSON, B. E., QUENSEL, C. E. LANKE, L. S., LUNDQVIST, C., GRAHNEN, H., BONOW, B. E. and KRASSE, B., *Acta odont. scand.*, **11**, 232 (1954)
10. KING, J. D., MELLANBY, M., STONES, H. H. and GREEN, H. N., *Spec. Rep. Ser. med. Res. Coun. No. 288* (1955)
11. GREEN, R. M. and HARTLES, R. L., *Br. J. Nutr.*, **21**, 225 (1967a)
12. GREEN, R. M. and HARTLES, R. L., *Br. J. Nutr.*, **21**, 921 (1967b)
13. GREEN, R. M. and HARTLES, R. L., *Br. J. Nutr.*, **20**, 317 (1966)
14. GREEN, R. M. and HARTLES, R. L., *Archs oral Biol.*, **14**, 235 (1969)
15. SHAW, J. H., KRUMINS, I. and GIBBONS, R. J., *Archs oral Biol.*, **12**, 755 (1967)
16. LEACH, S. A., *Adv. Fluor. Res. and dent. Caries Prev.*, **3**, 187 (1965)
17. GIBBONS, R. J. and SOCRANSKY, S. S., *Archs oral Biol.*, **7**, 73 (1962)
18. FITZGERALD, R. J., JORDAN, H. V. and STANLEY, H. R., *J. dent. Res.*, **39**, 923 (1960)
19. BERMAN, K. S. and GIBBONS, R. J., *Archs oral Biol.*, **11**, 533 (1966)
20. WEISS, S., KING, W. J., KESTENBAUM, R. C. and DONOHUE, J. J., *Ann. N.Y. Acad. Sci.*, **131**, 839 (1965)
21. WOOD, J. M. and CRITCHLEY, P., *Archs oral Biol.*, **11**, 1039 (1966)
22. GIBBONS, R. J. and BANGHART, S. B., *Archs oral Biol.*, **12**, 11 (1967)
23. WOOD, J. M., *Archs oral Biol.*, **12**, 849 (1967)
24. WOOD, J. M., *Archs oral Biol.*, **14**, 161 (1969)

25. GIBBONS, R. J., BERMAN, K. S., KNOETTNER, P. and KAPSIMALIS, B., *Archs oral Biol.,* **11,** 549 (1966)
26. WOOD, J. M., *J. dent. Res.,* **43,** 955 (1964)
27. KRASSE, B., *Archs oral Biol.,* **10,** 215 (1965a)
28. KRASSE, B., *Archs oral Biol.,* **10,** 223 (1965b)
29. HAYES, M. L. and LEACH, S. A., *J. dent. Res.,* **47,** 984 (1968)

Chapter Twenty

SUCROSE IN THE AETIOLOGY OF CORONARY THROMBOSIS AND OTHER DISEASES

JOHN YUDKIN
Professor of Nutrition and Dietetics, Queen Elizabeth College, University of London

There is evidence, of very varying strength, that a high consumption of sucrose is a contributory factor in the causation of several human diseases. Amongst these are dental caries, obesity, coronary thrombosis, diabetes, and certain diseases of the skin, the digestive tract and the joints. The role of sugar in dental caries is the subject of a separate paper in this conference *(Chapter 19)*. Before considering the evidence in relation to the other conditions, it may be useful to make some general observations on the type of evidence that is available.

1 EPIDEMIOLOGICAL EVIDENCE

If sugar is a cause of a specific disease, there should be an association between the disease and the consumption of sugar. In populations, we have seen that sugar intake is related to national incomes *(Chapter 1)*; the disease should therefore be more prevalent in wealthy countries than in poorer countries, and should have become increasingly prevalent as countries have become wealthy. There are however several problems involved in producing satisfactory epidemiological evidence. Firstly, statistics of the prevalence of particular diseases may be at worst quite unreliable, or at best uncertain. Relevant statistics do not exist for some of the poorer countries which one wishes to compare with the wealthy countries. For some diseases, statistics are not available even in the most advanced countries; for example no one knows precisely the prevalence of obesity, dental caries, diabetes or arthritis in the United Kingdom. Finally, even where statistics are kept most meticulously, as they are for example for mortality due to coronary thrombosis in this country, doubt may be cast on their accuracy on the ground of changing criteria of diagnosis by the doctors who certify causes of death.

In spite of these limitations, it is agreed by most workers that several of the diseases we have mentioned are indeed more common in wealthy countries than in poorer, and have increased in prevalence during the last fifty years or so with the increased living standards in the wealthy countries. It is for this reason that these diseases are sometimes referred to

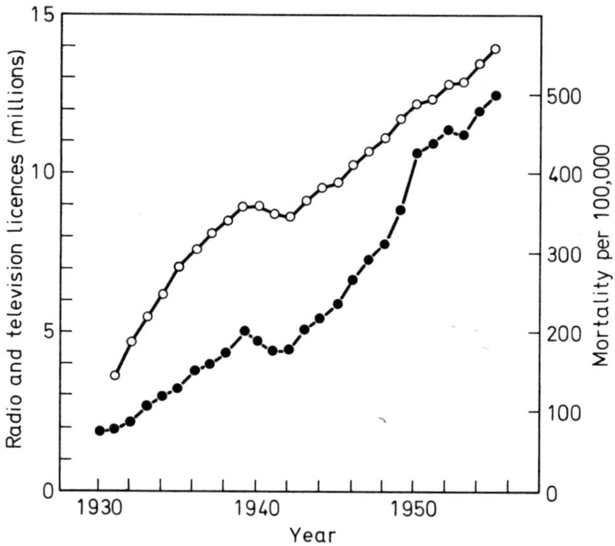

Figure 1. Trends in coronary mortality and number of radio and television licences in U.K.

as 'diseases of civilisation', although a preferable term would be 'diseases of affluence'.

A particular problem in relating diet to disease occurs if there is a long latent period before the disease becomes manifest; it is believed for example that maturity onset diabetes and coronary thrombosis may take twenty years or more to develop. It is difficult enough to assess current dietary intake; it is much more difficult to assess intake over a period of many years. In regard to sugar, we have especially the difficulty that, in many countries, consumption has been rising very rapidly over the past ten or twenty years.

It should be said here that the association between diet and disease does not prove causality. Some twelve years ago, I pointed out that the rise in recorded deaths due to coronary thrombosis in the United Kingdom was most closely related to the rise in the number of radio and television licences *(Figure 1)*. On the other hand, epidemiological evidence can be useful in eliminating some environmental factors as possible causes for disease. The white flour used in baking modern bread has been consumed in Britain for the last one hundred years, and the amount consumed has been falling steadily. It cannot possibly therefore be the cause of an increase in coronary disease.

2 EXPERIMENTAL EVIDENCE

The techniques of nutritional experiments with laboratory animals are now well established. Nevertheless, we often forget the limitations of the conclusions that we can reasonably draw from animal experiments. This is especially true for some of the diseases we have mentioned.

The most serious limitation is that some human diseases are not precisely reproduceable in animals, or at any rate not in the more common laboratory animals such as the rat, guinea-pig or rabbit. This is the situation with coronary thrombosis, although it can in fact be produced—laboriously and at great expense—in some of the higher apes.

A second limitation is, I believe, the failure to appreciate the different growth patterns of the experimental animals we use. The rat increases in weight for most of its three or four years of life; experiments on obesity that are made on rats, especially during their earlier rapid growth phase, cannot readily be interpreted in relation to human obesity, particularly if measurements are made only on body weight and not on the proportion of the major body constituents—water, protein and fat. Again, whilst it is true that all animal cells need the same components—amino acids, vitamins and minerals—in approximately the same proportions, different species of animals are adapted to quite different diets from which they obtain these components. The rabbit, commonly used in experiments on atherosclerosis, is often fed on diets rich in saturated fat and cholesterol. Since these are both quite foreign to the rabbit's normal diet, the way in which they are metabolised by the rabbit is unlikely to be identical to the way in which they are metabolised by omnivorous or carnivorous animals.

Dietary experiments may be carried out in human volunteers as well as in laboratory animals. This however raises other problems. One is the ethical problem; it would be reasonable to introduce a treatment that may reduce the chances of people developing coronary thrombosis, but not reasonable to introduce one that may increase the chances of developing the disease. A less difficult problem, but still a formidable one, is that of supervision of the diet and perhaps of physical activity, linked with the question of adequate controls. Overcoming these problems involves a budget that is beyond the reach of many research workers. One is thus often restricted to experiments with relatively few people, and over short periods of time. It would however be just as great a mistake to dismiss the results of such experiments as valueless because of these limitations, as to accept them uncritically as answering questions relating to long-term diets in all persons.

Finally, it should be said that evidence that dietary sugar is a cause of a disease does not imply that it is the sole cause. We know for example that cigarette smoking and physical inactivity are involved in the causation of coronary disease.

Let us now turn to the diseases of affluence to which we referred earlier, and examine the evidence that sugar consumption may be a causative factor.

The ingestion of large amounts of sucrose may be either at the expense of other foods, or in addition to other foods, or of course partly the one and partly the other.

1 NUTRITIONAL DEFICIENCY

In so far as sucrose may replace other foods in the diet, it will clearly reduce the intake of nutrients, since of all foods sucrose uniquely supplies calories and no nutrients. It is this property that has led to the description of sucrose as 'empty calories'. The average intake of sucrose in Britain contributes nearly 20% of all calories, and for many people it contributes substantially more.[1] Unless therefore the other foods making up the diet contain a relatively high proportion of nutrients, there is the inevitable hazard of nutritional deficiency. Although this is often commented on by nutritionists, I know of no published work that relates the intake of sugar to the intake of nutrients, but there is no doubt that the danger exists.

2 OBESITY

Recent research suggests that we must substantially revise some of the widely accepted views about obesity. It is not true

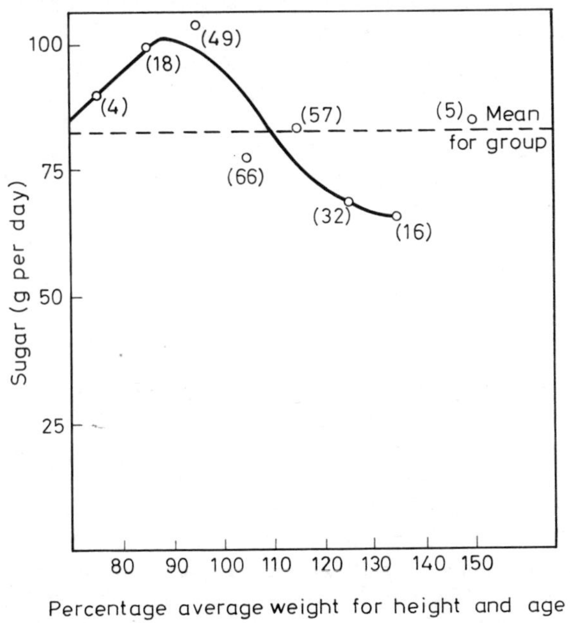

Figure 2. Sugar intake and body weight (figures in brackets refer to number of subjects)

that food providing more calories than are required by a person for his normal activity must inevitably be laid down as fat; some people have the facility to metabolise excessive food and to dissipate the resulting calories as heat.[2] Secondly, the same quantity of food when taken in several small meals in a day is less likely to produce obesity than when it is taken in one or two meals.[3] Thirdly, it seems as if some people are more likely to put on weight when the major carbohydrate source is sucrose rather than starch.[4] All of these reasons would reduce the chances of finding a close correlation between body weight and sugar consumption *(Figure 2)*.

Again, experiments in laboratory animals are complicated by the considerations mentioned earlier, that different species are likely to react differently to changes in the proportion of the major dietary components. Experiments with rats have suggested that obesity occurs when the diet is rich in fat. This would imply that the best diet for weight reduction in man would be one low in fat. Yet observations in obese human subjects suggest that the most satisfactory loss of weight occurs when the diet is low in carbohydrate.[5] Moreover, animal experiments in which sugar is given instead of starch produces equivocal results; depending on the sex of the animal, its age, and the duration of the experiment, sugar may cause an increase or a decrease in body fat, or no change.[6] All we can say is that from clinical experience we know that many overweight people lose weight simply by considerable reduction in sugar intake.

3 DENTAL CARIES

There is a considerable literature on the role of sucrose in the production of dental caries. We ourselves have carried out a few investigations both with rats and with children. With rats, we have confirmed that dietary sucrose produces much more dental caries than does dietary starch, and that the amount of caries is proportional to the amount of sucrose in a mixture of both these carbohydrates.[7] In children, we have shown that there is a relationship between the DMF index (the number of decayed, missing or filled teeth) and the amount of sugar taken in solid foods *(Table 1)*. This relationship was, as expected, strongest in children who admitted to poor dental hygiene, but disappeared in those who claimed to clean their teeth regularly *(Table 2)*.

4 CORONARY HEART DISEASE

Epidemiological evidence from population studies shows a close correlation between sugar consumption and the preva-

Table 1. INTAKE OF SUGAR IN SOLID FOODS, AND DENTAL CARIES IN BOYS AND GIRLS AGED 15–18

	Boys	*Girls*
Number	89	44
Correlation coefficient	0·24	0·25
P value	<0·02	<0·05
DMF Index of high consumers	11·8	15·9
DMF Index of low consumers	9·0	12·3

Table 2. CORRELATION BETWEEN CARIES AND SUGAR INTAKE, IN RELATION TO DENTAL HYGIENE IN GIRLS AGED 15–18

Dental hygiene	*Number*	*P value*
Poor	12	0·001 (r = 0·81)
Average	12	NS
Good	19	NS

Table 3. SUGAR INTAKE OF SUBJECTS WITH MYOCARDIAL INFARCTION (MI) OR PERIPHERAL ARTERIAL DISEASE (PA), AND OF CONTROL SUBJECTS (C). MEN AGED 45–65

Group	*Number*	*Age* (*mean*, years)	*Sugar intake* (g per day)	
			Mean	*Median*
MI	20	56·4	132	113
PA	25	56·5	141	128
C	25	56·0	77	56

Significance of differences in sugar intake: MI and control: $P<0{\cdot}0005$
PA and control: $P<0{\cdot}00003$

Table 4. SUGAR INTAKE OF SUBJECTS WITH MYOCARDIAL INFARCTION (MI) AND OF CONTROL SUBJECTS (C2 HEALTHY MEN FROM FACTORY, C1 HOSPITAL PATIENTS). MEN AGED 45–63

Group	*Number*	*Age*, years		*Sugar intake* (g per day)	
		Mean	*Median*	*Mean*	*Median*
MI	20	55·4	57	148	147
C2	20	55·4	57	78	67
C1	13	53·1	50	90	74

Significance of difference in sugar intake: MI and C2: $P<0{\cdot}001$.
MI and C1: $P<0{\cdot}01$.

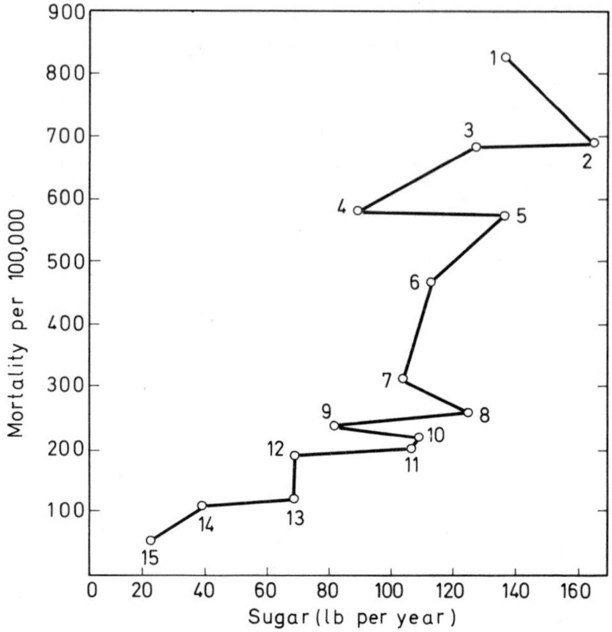

Figure 3. Coronary mortality and sugar intake in various countries

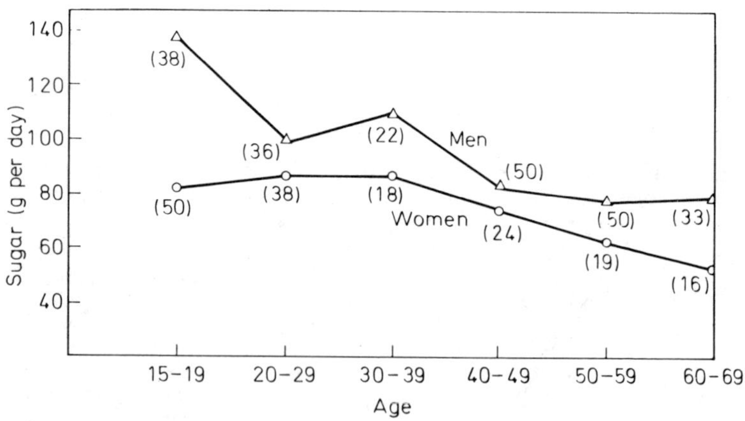

Figure 4. Sugar intake, age and sex (figures in brackets refer to number of subjects)

lence of the disease[8] *(Figure 3)*. In addition, several studies have shown that, in a given population, individuals who develop coronary thrombosis have been eating more sugar than individuals who do not develop the disease[9, 10] *(Tables 3, 4)*. Because of variation in sugar intake according to sex,

Table 5. EFFECT OF DIETARY SUCROSE ON INSULIN RESPONSE IN NINETEEN SUBJECTS (MEAN, μU/ml)

Dietary period	*Thirteen subjects* 0 min	30 min	120 min	*Six subjects* 0 min	30 min	120 min
Preliminary	22	63	33	23	64	34
Sucrose	23	65	33	39*	131*	84*
Normal	22	65	34	25	72	40
Starch	22	64	33	22	64	32
Normal	21	64	31	23	63	34

*Significant difference between these values and those for preceding and subsequent period ($P<0{\cdot}01$)

age, class and nationality, it is essential in such studies to match very carefully the two groups of individuals *(Figure 4)*.

Experimental evidence is mostly indirect, consisting of inducing the sorts of changes in animals or volunteers that accompany those seen in coronary disease. Most of the work

Table 6. CHANGES IN PLATELET ADHESIVENESS (%) IN SIX SUBJECTS WHO DEVELOPED HYPERINSULINISM AND THIRTEEN WHO DID NOT

	Dietary period				
	Preliminary	*Sucrose*	*Normal*	*Starch*	*Normal*
Thirteen subjects	67	67	68	69	66
Six subjects	68	60*	71	65	68

*Significant difference between this value and those for preceding and subsequent period ($P<0{\cdot}05$)

reported hitherto has been on the level of cholesterol in the blood, which is more likely to be raised in those who are prone to develop coronary disease than in those who are not. More recently it has been suggested that there is a closer relationship with the level of triglyceride.[11] With a change in dietary carbohydrate from starch to sucrose in experimental animals, many workers have found a rise in one or both of these lipids. But there is now reason to believe that there are more relevant changes in the blood of coronary-prone individuals than these lipid changes; notably, attention has been drawn to changes in carbohydrate metabolism, in the blood clotting mechanism, and in the level of insulin in the blood. In this connection, we have found that about one-third of a group of men given a diet rich in sugar showed an increase in insulin and in the tendency of the blood platelets to stick together, a phenomenon that probably initiates the blood-clotting process[4] *(Tables 5, 6)*.

5 DIABETES

Diabetes that occurs in middle age, as distinct from juvenile diabetes, is more frequently seen in wealthy countries. The evidence again points to an association with sucrose consumption.[1] Especially, it has been shown that both experimental animals and human subjects develop an impairment in glucose tolerance when fed for several weeks on diets rich in sucrose.[12, 13] It is perhaps relevant to point out here that there is an association between maturity onset diabetes and coronary heart disease; patients who show symptoms of the one disease very frequently have, or develop, manifestations of the other disease.

6 DISEASES OF THE SKIN, JOINTS AND DIGESTIVE TRACT

There is much less evidence relating sugar consumption to these diseases than that relating it to the diseases discussed so far. Seborrhoeic dermatitis occurs in people who consume significantly more sugar than matched controls[14] *(Table 7)*. A therapeutic trial is now under way to see the effect upon this skin disease of reducing the amount of sugar in the diet. Again, there are indications that patients with rheumatoid arthritis and with gout also consume a higher amount of sugar than control subjects.

As regards digestive disease, the evidence is even more tenuous. Nevertheless, there are indications that chronic dyspepsia, gastric and duodenal ulcers, and gall bladder disease may be affected by a high intake of sugar. Those of us who have been treating obese patients with the low-carbohydrate diet have observed the tremendous relief from pain and discomfort that patients report almost from the moment that they reduce their carbohydrate intake. This is

Table 7. SUGAR INTAKE OF SUBJECTS WITH SEBORRHOEIC DERMATITIS AND OF CONTROL SUBJECTS (CONTROL 1: PATIENTS WITH WARTS; CONTROL 2: HEALTHY FACTORY AND OFFICE WORKERS)

Group	*No.*	*Age (Mean, years)*	*Sugar intake* (g/day)			
			Total		*Beverage*	
			Mean	*Median*	*Mean*	*Median*
Dermatitis	16	38·2	170	185	104	109
Control 1	13	36·5	108	109	65	56
Control 2	16	38·1	110	110	61	57

Significant differences.—Total sugar intake: dermatitis and control 1—$P<0{\cdot}01$. Dermatitis and control 2—$P<0{\cdot}02$. Beverage sugar.—Dermatitis and control 1—$P<0{\cdot}02$. Dermatitis and control 2—$P<0{\cdot}05$.

particularly true of the reduction of dietary sugar. As with seborrhoeic dermatitis, more solid evidence may be forthcoming when a current therapeutic test on a low-sugar diet in patients with these diseases is completed.

THE MECHANISM OF ACTION OF SUCROSE

One reason that has led people to doubt the intervention of sucrose in the causation of these diseases was the apparent absence of any suggested mechanism whereby sucrose might cause its effect. We are still far from understanding in detail what these mechanisms can be for all these diseases, except that it is already evident that they are likely to be manifold. It seems certain that sucrose promotes dental decay because of its unique facility of forming the dextran upon which the cariogenic organisms flourish. We believe that the role of sucrose in causing diseases of the digestive tract derives from the fact that, in the large amounts we now consume it, it is the only dietary component that contributes to the creation of a high osmotic pressure, and hence to chronic irritation of the gastric and duodenal mucosa.

As to the role of sugar in the causation of coronary disease and diabetes, we are now beginning to believe that an early effect, if not the earliest effect, is the rise in circulating insulin. From this, we can dimly perceive a picture of alteration in other hormones, and in enzyme patterns that determine metabolic rates and even metabolic pathways, which could well lead to the large complex of changes that underly both of these diseases. Much more work is needed however before we can answer all the questions that are raised by these considerations.

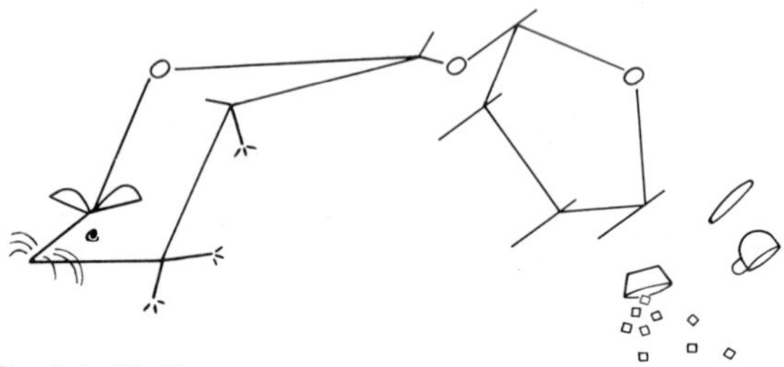

Design by Patricia Yudkin

REFERENCES

1. YUDKIN, J., *Proc. Nutr. Soc.*, **23,** 149 (1964)
2. MILLER, D. S., MUMFORD, P. M. and STOCK, M. J., *Am. J. clin. Nutr.,* **20,** 1223 (1967)
3. FABRY, P. and BRAUN, T., *Proc. Nutr. Soc.*, **26,** 144 (1967)
4. SZANTO, S. and YUDKIN, J., *Postgrad, med. J.*, **45,** 605 (1969)
5. YUDKIN, J., *This Slimming Business,* Macgibbon and Kee, London (1958)
6. AL-NAGDY, S., MILLER, D. S. and YUDKIN, J., *Nutrition and Metabolism,* in the press (1970)
7. SALTER, A. J. and GRENBY, J. H., *Caries Research,* **3,** 200 (1969)
8. YUDKIN, J., *Lancet,* **II,** 155 (1957)
9. YUDKIN, J. and RODDY, J., *Lancet,* **II,** 6 (1964)
10. YUDKIN, J. and MORLAND, J., *Am. J. clin. Nutr.*, **20,** 503 (1967)
11. ALBRINK, M. J., MEIGS, J. W. and MAN, E. B., *Am. J. Med.,* **31,** 4 (1961)
12. COHEN, A. M. and TEITELBAUM, A., *Am. J. Physiol.,* **206,** 105 (1964)
13. COHEN, A. M., TEITELBAUM, A., BALOGH, M. and GROEN, J. J., *Am. J. clin. Nutr.,* **19,** 59 (1966)
14. BETT, D. G. G., MORLAND, J. and YUDKIN, J., *Br. med. J.*, 3, 153 (1967)

INDEX